# Transition to Digital Ethics

As digital technologies become more pervasive and powerful, they raise increasingly pressing ethical questions that need to be understood and addressed.

This book provides an accessible introduction to the important field of digital ethics. It explores topical examples of ethical problems created by the technology and shows how ethical ideas and values can be used to make sense of them. It provides the readership a clear, well-researched insight into the various issues under the banner of 'digital ethics', including how each of these issues relates to different technologies such as AI, smartphones, and automated decision-making (to say the least). This is coupled with contemporary ideas on ameliorating ethical harm and risk, which are grounded in philosophical thought, presented in an accessible way.

The book will benefit those who design and implement digital technologies and, more broadly, anyone with an interest in the compelling ethical and social issues arising from digital technologies. It carefully and clearly describes the philosophical ideas as well as technical considerations behind contemporary digital technology use, to provide the reader with a solid footing for further exploration of this rapidly expanding field.

**Marc Cheong** is Deputy Director of the Centre for AI and Digital Ethics (CAIDE) and Senior Lecturer of Information Systems (Digital Ethics) at the University of Melbourne, and Honorary Burnet Institute Senior Fellow. His research interests are in social media, bridging philosophy/ethics and IS, and societal implications of technology, among others.

**Simon Coghlan** is Deputy Director of CAIDE and Senior Lecturer of Digital Ethics at the University of Melbourne. He is a moral philosopher and a former veterinarian. His research interests are in AI ethics, including how AI affects humans, animals, and the environment, and in robot ethics, including the ethics of human interactions and relations with machines.

# Transition to Digital Ethics

## A Primer from Philosophy to Practice

Marc Cheong and Simon Coghlan

CRC Press
Taylor & Francis Group
Boca Raton London New York

CRC Press is an imprint of the
Taylor & Francis Group, an **informa** business
A CHAPMAN & HALL BOOK

Designed cover image: Shutterstock

First edition published 2026
by CRC Press
2385 NW Executive Center Drive, Suite 320, Boca Raton FL 33431

and by CRC Press
4 Park Square, Milton Park, Abingdon, Oxon, OX14 4RN

*CRC Press is an imprint of Taylor & Francis Group, LLC*



ISBN: 9781032493824 (hbk)
ISBN: 9781032493848 (pbk)
ISBN: 9781003393511 (ebk)

DOI: 10.1201/9781003393511

Typeset in Palatino
by KnowledgeWorks Global Ltd.

## Dedication

*MC: This book is dedicated in loving memory of Judy Bastone (1941–2023) and Patrick 'P.J.' Byrne (1956–2017). Your company, kindness, and wisdom will be forever missed.*

*SC: This book is dedicated to my family, Sarah, Charlotte, and Will, for their loving support.*

# *Contents*

# *Acknowledgments*

MC and SC:

We would like to first and foremost acknowledge the generous help of (soon-to-be-Dr) Susie Sheldrick for being our eagle-eyed copyeditor and research assistant for this volume. Without Susie's constant motivation, positive feedback, and collegiality, we would not be able to finish this book. Thank you, Susie (and Basil)!

To our Editorial team at Taylor & Francis/Chapman & Hall, especially Gerald Bok, Payal Bharti, and George Knott: Thank you for your help in making this idea a reality.

We would like to express our heartfelt gratitude to CAIDE: especially Jeannie Paterson and Abi Ward for their invaluable contributions and support, not forgetting Tim Miller and Gabby Bush for the ideas behind this book.

MC:

I would like to thank my family – Jo Byrne, Henry Cheong, Rose Cheah, Diane Byrne – for being so patient with me, and for giving me the confidence to finish this book.

To my generous colleagues (too many to list), thank you for all your ideas, wisdom, and input, which I appreciate and duly acknowledge in the footnotes to this book.

To those whom I have solicited input for this book – Kean, Jolene, Chong and Elise; Kevin, Yik, and Kien-Seng; and Imbi; among others – *Marky Marc* thanks you for your contributions.

To all my mates, fellow Aussie quizzers, and the Australian Quokkas quizzing team: Thanks for having me as part of the team.

And most importantly, to Simon, my co-author – thanks for embarking on this journey with me. I couldn't think of a better co-author for this project.

# Preface

## The Explosion of Digital Technology

Digital technology is now evolving at an astonishing rate. This acceleration in digital technology is powered by advancements in artificial intelligence (AI), software development, more powerful hardware, and a growing appetite for newer and better things. The sheer speed, power, and reach of digital technology can be tremendously exciting. We are entering a world in which computers can converse with us, write essays and jokes, and produce images in any style.

Digital systems are increasingly being entrusted to form predictions about the world and to make decisions that affect our employment and wellbeing. Robots are likely to become more able to sense, navigate, and interact with us. Machines and digital avatars may even become our personal assistants and 'companions'.

While exciting, this rapid change also gives rise to pressing ethical concerns – concerns about what we should do with respect to digital technology. Some of these concerns may seem more like science fiction. For example, some experts predict the imminent arrival of AI with general intelligence like humans, perhaps to be followed by an explosion of ever more 'superintelligent' digital entities that could deliberately or mistakenly wipe out humanity or even life on earth. This obviously raises the question of whether we should be so enthusiastically building smarter and smarter digital systems.

But there are a great many more 'prosaic', if still critically important, ethical issues that are already arising or have arisen. Take the seemingly mundane example of a patient's medical file. Not all that long ago, sensitive patient data was routinely kept on pieces of paper in manila files in doctors' surgeries. With the advent of digitalisation, the data could be stored on a computer, enabling rapid processing and searching of information. Before the internet, this data could hypothetically still be leaked or stolen by never-do-gooders, but only if they had physical access to the surgery's computer.

Nowadays, electronic medical records (EMRs) are stored on servers connected via networks and the wider Internet. In some cases, data from different medical providers about a single patient can be stored together in aggregate. If a breach were to happen, the damage may be much less localised. In fact, the entire medical history of the patient – and of many other patients – can be stolen *en masse*. This raises ethical and legal questions about what measures are required to protect private patient data.

Let's take a broader and quick trip down memory lane to set up the scene for the rest of this book. A significant acceleration in digital technology took place at the turn of the third millennium. To put things in context, consider the state of technology approximately 20 years ago. Self-driving cars were a futuristic possibility. The world was still making the leap from landlines to mobile technologies. The short message service (SMS) and older technologies such as MSN Messenger were the 'in' way to communicate.

A quarter century ago, video calls were largely outside the reach of the home user. Online social networks were in their infancy, before Facebook became the world leader by owning several platforms. Multimedia sharing via peer-to-peer platforms depended on an up-and-coming platform called BitTorrent (but murky in terms of copyright): Hubs with legal access to high-quality movies and music were known as video and record stores. You could find information you needed on Google, which had just introduced a 'beta' version of its Gmail client.

Fast forward to the time of writing. Online communications platforms from Facebook Messenger to WhatsApp to Signal have superseded the good old SMS. The world now often depends on these platforms, as well as on household staples such as Zoom and FaceTime. This dependency grew and was dramatically shown during the COVID-19 pandemic.

Online social media platforms have come and gone: The 'Millennial' Vine (in the early-mid 2010s)[1] has been overtaken by the 'Gen Z' TikTok. With these products has come a whole suite of trends and concerns. These range from addiction to privacy loss, scams and financial fraud, and opportunities for crime and radicalisation. Streaming media is the rage nowadays, with a myriad of video and audio streaming platforms, all in time to keep us company during pandemic lockdowns and help us catch up with life post-pandemic.

Various Generative AI systems, from the popular (Chat)GPT, to its competitors such as Mistral, DeepSeek, and Gemini, have moved to the fore since the end of 2022. Generative AI has found its way into our daily consumer products, both the tangible kind (smartphones) and the intangible kind (our daily email application). Self-driving cars are on some roads in the USA, although they have unfortunately claimed lives due to technical and other issues, which has slowed their wider acceptance.

Technological giants such as the *FAANG* big tech companies (an acronym for Facebook/Meta, Apple, Amazon, Netflix, and Google), and companies which enjoyed the AI boom such as Nvidia are everywhere. These tech titans have created concerns about companies and powerful individuals 'knowing too much' about us, including our search histories, video watching habits, who we socialise with, where we have been, what our preferences are. This data is then used by those entities to predict what we might do and think. Increasingly, people are concerned about digital platforms like social apps targeting them with advertisements for things they mentioned in person to their friends, even when they are not actively using the app![2]

Where once we had a range of 'non-digital' ethical issues to contend with, we now have also to worry about ethical issues stemming from our use and appreciation of, and indeed our growing dependency on, digital technology. The continual advancement in AI, the mounting quantities of data being generated, the booming of digital platforms and systems, and the subsequent potential for our lives to be impacted, can create unease and confusion. We all now have a stake in ethical reflection in these extraordinary, and often murky and uncertain, times.

## Aims of Book and Intended Audience

The authors of this book have taught and briefed various people over the years in digital ethics and related subjects: ranging from organisational and governmental decision-makers; to undergraduate, postgraduate, and doctoral students; to school students and general members of the public.

We understand the challenges faced by learners of digital ethics wanting to broach the topic for the first time. Newbies to digital ethics can face pedagogical constraints (grasping philosophical foundations), interdisciplinary barriers (especially where technologists and humanities scholars lack a shared vocabulary and understanding), and practical concerns (*I just don't have the time!* is a common, and very understandable, reply).

In the spirit of the *Transition to Tech* series, this book is aimed at 'people moving from other sectors into technology or seeking to use technology in their own sector'. In other words, we do not expect you to know about the inner workings of AI, nor about philosophy, ethics, or law.

This book is written with that in mind. It seeks to inform a general, professional, audience who has an interest in technology and its effects on our own lives, the people around us, wider society, and the environment.

Therefore, our book has the following aims:

1. Introduce you, the reader, to the current issues facing technology globally, and link them to ethical concerns in the area we term 'digital ethics'.
2. Provide an introduction to ethics and philosophy, in a *practical, applied, and accessible manner.* We aim to provide you with working knowledge on key theories, without overwhelming the reader: We are not intending this to have the same breadth and depth as, say, a full university course on philosophy and ethics, or a larger textbook.

3. Walk you through several case studies and examples of issues found in today's usage of technology and *applying what we have learnt* in the process. Digital ethics is like a sport or exercising at the gym – practice helps.
4. Acknowledge the *limitations* of ethics in that it cannot practically solve everything that can go wrong in our use of technology. To borrow a common maxim: It is no silver bullet. We cannot expect ethics to solve all our problems, since we need to take a range of other actions.
5. Provide ideas for *future explorations* on the subject, and make some observations on the current evolution of technology and what it means to us.

We hope you are as excited in this journey as we are in taking you on a guided tour, which is our *Primer from Philosophy to Practice.*

## Avoiding the Twin Vices of Panic and Complacency

The spirit of this book – including the case studies, discussions, and provocations presented within – is a tempered, considered, and balanced introduction to the topic. We need, first and foremost, to avoid what we term the 'twin vices of panic and complacency'.

What does this mean? As we shall see later in Chapter 4, philosophically speaking,[3] a vice is a human tendency and disposition of character, which is flawed.

Both panic and complacency can be problematic. Consider having to do a 10 km running race. For most of us, it is obviously necessary to plan the exercise regime, build up stamina, condition for various running speeds, and so forth. It is no good being so complacent as to leave training to the last minute. It is also of no use to panic and start running extreme distances from the get-go, without proper slow and steady conditioning. Either option could ruin our race.

The same goes for our approach to digital technology. Being ethically complacent could cause us to overlook critical issues. Or it could result in being stuck in a corner if, or rather when, things go wrong. Complacency may deprive us of the moral concepts and tools required to understand and address such problems. Furthermore, if we fail to anticipate and prepare for ethical failures, we may end up investing more time and resources to address harms after the horse has bolted.

Sometimes these harms can, sadly, be impossible to fully fix or repair. Take the Australian Robodebt scandal that emerged from the late 2010s. In this salutary example, an improperly implemented automated digital system was used to recoup Government debt from people on welfare. Unfortunately, it caused irreparable harm to many individuals.[4] As indicated in the final report from the Royal Commission into Robodebt, the scheme was a crude and cruel mechanism, neither fair nor legal, and it made many people feel like criminals. Many innocent people were deeply traumatised; some took their lives. Robodebt was a costly failure of public administration, in both human and economic terms.[5] Moral complacency about the harms of this automated system led to human tragedy.

On the other hand, panic about digital technology is not suitable either. Panic too deprives us of a careful, measured ethical approach, and is less likely to lead to appropriate and lasting change. If we lose our heads, we might also miss out on positive opportunities technology can bring. Sometimes we can be too quick to condemn technology, just as we can be too quick to embrace it. For example, condemnation and criticisms were swift when the everyday telephone was first introduced. According to Carolyn Marvin, these critiques ranged from privacy concerns (the telephone's 'potential to expose private family secrets') to the changing of societal behaviours ('essential markers of social distance were in danger'), which led to an interesting observation:

> Energetic efforts were made, therefore, to limit opportunities for new instruments of communication to create new secrets, and to protect the old ones they put at risk.[6]

Suffice to say that despite these anachronistic concerns, the telephone is still with us today, albeit with traditional landlines being replaced in favour of mobiles.

We are glad that you're picking up this book, as it is the first step to understanding the main ethical issues and challenges around technology. We believe, with a balanced approach to understanding the issues, their implications, and how we can remedy them at various stages, you the reader be will on track to think like a digital ethicist.

## Notes

1. See https://en.wikipedia.org/wiki/Vine_(service).
2. In reality, this might be due to the precision in which social media platforms can 'microtarget' us and the people who are around us, rather than surreptitiously recording our conversations or spying on us, contrary to the popular theory floating online (see, e.g., Chapter 2).
3. Throughout this book, we may introduce new philosophical/ethical/technical terms that might be new to the reader, or are different uses of existing common words in the everyday. These words will be denoted in **bold**.

4. Murray, Cheong, & Paterson (2023, July 10).
5. Commonwealth of Australia (2023).
6. Marvin (1990).

## References


Commonwealth of Australia. (2023). Royal Commission into the Robodebt Scheme (Vol. 1). https://robodebt.royalcommission.gov.au/system/files/2023-09/rrc-accessible-full-report.PDF

Marvin, C. (1990). When Old Technologies Were New: Thinking About Electric Communication in the Late Nineteenth Century. Oxford University Press. https://doi.org/10.1093/oso/9780195063417.001.0001

Murray, T., Cheong, M., & Paterson, J. (2023, July 10). The Flawed Algorithm at the Heart of Robodebt. Pursuit – The University of Melbourne. https://pursuit.unimelb.edu.au/articles/the-flawed-algorithm-at-the-heart-of-robodebt

# 1

# *Introduction*

## Why Do We Need Digital Ethics?

Before we ask the question, "why digital ethics?", let's start with a more basic one: "what is ethics anyway?"

The Oxford Dictionary defines ethics as:

> … moral principles that govern a person's behaviour or the conducting of an activity.[1]

These 'principles' about right and wrong behaviour or action could be well thought out, or they might be moral values and rules that we act on without much reflection. In any case, we all have ethical principles and values that we have internalised and that we live by. And this means more than abiding by the law or following other people's commands. It also means more than not just pursuing our own self-interest. For ethics additionally implies thinking about how our behaviour affects the interests of others.

The next question we could ask is, *what makes digital ethics special*?

Before we answer that, consider what has made the field of biomedical ethics special. There have been many extraordinary medical advancements in the past century or so, from modern surgery to imaging and vaccines. Yet, because these powerful medical treatments and procedures apply to real human beings, who may be either greatly helped or seriously harmed by health interventions, there needs to be substantial ethical thought about what medical practitioners ought to do. For instance, if a procedure such as a blood transfusion could prevent a person's imminent death, is it ethical for a medical professional to administer it if that person refuses? Is it right to allow doctor-assisted euthanasia for patients suffering with various diseases who wish to die, even though modern treatment might keep them alive for much longer?

Medicine contains many more hard ethical questions. To address these questions, we need to take time to think them through and discuss them with others. We also now know about various scandalous practices[2] in the history of medical research which might have been avoided had well-established principles and codes of medical ethics been in place beforehand. Though ethical principles and codes cannot fully stop people behaving

DOI: 10.1201/9781003393511-1 

unethically, they can at least have some positive effects, including increasing our understanding.

It is in this spirit that we are writing this book. Advancements in digital technology, which we encountered in the Preface, have been accelerating over time. Lately, researchers and even the general public have become more aware of the tricky moral questions and the ethical pitfalls and failures of such technology. With the speed these technologies now 'come to market' and spread like viruses – recall ChatGPT's tremendous early growth – it is vital that designers, deployers, policymakers, and users gain an awareness of digital ethics.

Even so, a sceptic of digital ethics might ask: *Hang on, I don't really use that much AI or fancy tech or even social media, so how does it affect me?* That is for some people a natural response. However, the sceptic might consider their newest mobile phone purchase, for instance. Are they aware of the terms and conditions that they agree to when they use a mobile phone? These might include things like the right to collect and use personal data – one's name, contact information, and location data – and the right to serve people targeted advertising based on their activity. Consider also the 'cookie popups' on the websites that online users frequent (see also Chapter 2). While these popups are intended to inform users about data tracking and gather consent, they often have intrusive and confusing designs, which can create frustration and friction, encouraging users to hastily accept tracking or simply to leave the site altogether. This behaviour from online platforms can compromise ethical principles of informed consent and transparency.

Also, suppose a person really does not use any new technology, but sticks to reading books (like this one) and watching 'old school' movies and television. Surely digital ethics does not apply here?. But now we must confess that most of the previous paragraph was not written by a human but was rather generated using state-of-the-art large language models. (*We can assure you that the rest of the words in this tome are written by humans, but hopefully this trick is still fresh enough to make readers pause*). And what if that documentary you have been watching, unbeknownst to you, used deepfake technology to 'reincarnate' a deceased celebrity, as infamously happened in the case of Anthony Bourdain and his artificially manufactured voice?[3] All these, and more, are ethical issues we are concerned about in digital ethics.

Now suppose a reader, although being interested in technology, nonetheless has this response: *I am not a philosopher, so I don't think I can answer these ethical questions about tech.*

In reply to this worry, we shall borrow a thought from philosopher Matthew Chrisman and his colleagues. They explain in their aptly titled *Philosophy for Everyone* that everyone, including non-philosophers, may turn to philosophy:

> ... to unbar the intellectual holds ... to help [us] begin to think reflectively about issues that we all already, in some implicit and unreflective sense, know our way around.[4]

We all have some implicit knowledge and understanding of ethics from ordinary life. But we also probably have had some exposure to issues in digital ethics. The digital technologies we use – from mobile phones to social media – and the digital technologies we are affected by – as when we read an AI-generated piece of work or doubt the online media we consume – give rise to issues we already 'know our way around', at least to some extent. Having some philosophical tools for thinking can help us to deepen this understanding.

Technology does not exist independently of individuals, communities, cultures, and political systems. Rather, technology and society are intertwined and can influence each other. The example of the problems of, say, discriminatory decision-making administrative tools, or social media, are striking illustrations. Social media shows how seemingly small design decisions in algorithms and interfaces, such as the creation of *Like* buttons and endless scrolling, can profoundly interact with human behaviours and social conditions. Interaction between technology and people can sometimes even change society and the way we live, as we have seen with social media, and as we will increasingly see with AI decision-making and AI that generates remarkably human-like text and astonishing images. The fact that digital technologies are growing increasingly complicated, pervasive, and entangled in society means that digital ethics is an increasingly large – and interesting! – branch of ethics.

## How Is Digital Ethics Related to Computer Science?

Computer science spans many subfields, from programming to the design of search engines like Google. There are at least ten common subfields of computer science, and each of these subfields can give rise to ethical issues.[5] We briefly introduce these forms of computer science here.

### Artificial Intelligence (AI)

While historically there is no one single definition of Artificial Intelligence or AI,[6] we can start with a simple definition from the OECD: AI is "... a machine-based system that ... infers, from the input it receives, how to generate outputs such as predictions, content, recommendations, or decisions".[7] Major gains in the power of AI have come about more recently due to improvements in *machine learning (ML)* (see below).

While AI has been around for decades, public and scholarly discourse on AI is gaining popularity thanks to the advent of Generative Artificial Intelligence or **GenAI**. GenAI is defined by McKinsey as AI algorithms ...

> ... that can be used to create new content, including audio, code, images, text, simulations, and videos.[8]

*Example of an ethical issue in this area:* How might GenAI cause harm when, during training, it learns from biases or historical stereotypes and recreates these biases or stereotypes in the output it generates?

### Computer Science (CS)

Computer science, or CS, has been through various changes in the past few decades. An early description of CS is "a science of abstraction – creating the right model for thinking about a problem and devising the appropriate mechani[s]able techniques to solve it".[9] As computers get more complex, we can simply define CS as "... the study of the principles and use of computers".[10]

*Example of an issue in this area:* How could a 'decision tree' algorithm (a collection of decision rules that 'branch out' akin to a flowchart) be made to optimise the distribution of resources, and to benefit data subjects fairly, in its decision-making?

### Computer Security (or Cybersecurity or Information Security, Abbreviated InfoSec)

Computer security is fast becoming necessary due to the complexities of the technology we now use every day. Take one's activity on an online banking website for example; if there is a loophole in the device we use, or the operating system, or the web browser, or the bank's web app itself, it could be exploited by malicious parties to illegally steal someone's funds. The increasing number of scams and fraud cases as well as cyberthreats directly impact one's privacy and sense of security.

*Example of an issue in this area:* How do we ensure users' personal information is secured from harm, such as identity theft and cyber-extortion?

### Computer Vision

Computer Vision (CV), which is related to CS and AI, "focuses on enabling computers to identify and understand objects and people in images and videos ...", and "derive meaningful information ... to make recommendations or take actions".[11] A popular use of CV is to recognise your face to unlock your device, or to detect faces from photos on your mobile phone (and automatically label people for faster photo searching), or to check up on pets' wellbeing (and even dispense treats!) when their owners are at work (Figure 1.1).

*Example of an issue in this area:* How can we make sure that algorithms for facial recognition are safe, non-discriminatory, and not misused by third parties for unjustified surveillance?

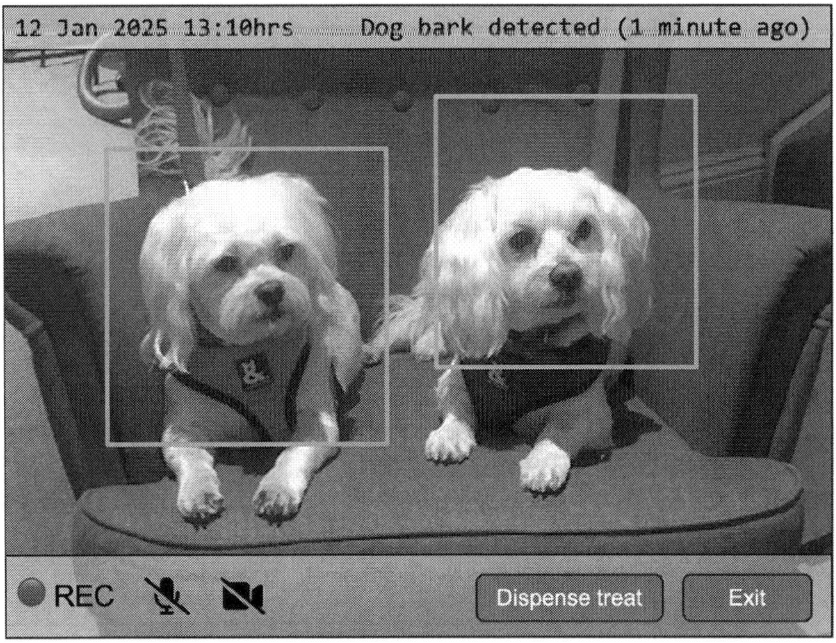


**FIGURE 1.1**
A typical example of a pet camera/feeder system, which notifies owners if it detects that pets are in distress, and allows owners to talk to their pets and dispense treats. (*Screen mockup by the authors, with icons from Font Awesome by Dave Gandy – http://fontawesome.io – under the SIL Open Font License*).

## Human–Computer Interaction (HCI)

As with, say, AI, there is a variety of definitions for human–computer interaction, or HCI. Simply put, HCI studies how humans interact with, use, and accomplish tasks using computers and related technologies. Areas studied in HCI range from ensuring accessibility and usability of interfaces; to research on principles, theories, and models guiding the development of user-facing systems; to prototyping and evaluation of new and proposed applications.[12]

*Example of an issue in this area:* How can we ensure that new technological artefacts we produce can be equitably used by people with different accessibility requirements?

## Information Systems

The Information Systems (IS) subfield looks at how a digital system "collects, processes, stores, analy[s]es, and disseminates" information "to the right people, at the right time, in the right amount, and in the right format".[13] Topics in IS range from business processes, to social media, to supply chains.

*Example of an issue in this area:* How do we ensure that organisational and community-oriented IS – from library databases to e-commerce sites – benefit their stakeholders?

### Machine Learning

ML is considered a type of AI and a subfield of CS "concerned with building algorithms which, to be useful, rely on a collection of examples of some phenomenon" – or, to be exact – "the process of solving a practical problem by (1) gathering a dataset, and (2) algorithmically building a statistical model based on that dataset ... to solve the practical problem".[14] ML lies behind many recent advances in AI.

*Example of an issue in this area:* How do we investigate how an ML algorithm used to rank candidates for a particular job is being fair to people of diverse ethnicities, genders, ages, and walks of life?

### Natural Language Processing (NLP)

Natural Language Processing (NLP) is an area related to AI and ML with the focus on enabling computers to comprehend, generate, manipulate, understand, and communicate with human language.[15]

*Example of an issue in this area:* How can we develop safer moderation techniques to stop cyberbullying and abusive language on social media platforms?

### Systems Architecture (SA)

Related to IS and Software Engineering (more later), Systems Architecture (or SA) deals, in part, with "the fundamental organization of a system, embodied in its components, their relationships to each other and the environment, and the principles governing its design and evolution".[16]

*Example of an issue in this area:* How can we build systems that support legal requirements which serve to protect stakeholders?

### Software Engineering (SE)

The textbook definition of Software Engineering (SE) is "an engineering discipline that is concerned with all aspects of software production from initial conception to operation and maintenance".[17] This discipline helps us make sense of how to build complex software applications that require huge teams of developers using tried-and-tested methods.

*Example of an issue in this area:* How could we have an inclusive community that empowers coders from the Global South to contribute to software development (e.g., the Open Source community working on the ubiquitously used Linux operating system)?

### Robotics

The aforementioned ten subfields broadly belong to CS, but we should not discount the field of robotics from our discussion on ethics. Robotics is often studied in branches of engineering, such as mechanical engineering, but it is also studied in fields such as HCI, and requires technologies from, say, CV and AI. Broadly speaking, a robot is a machine that can move around or has moving parts and can act (semi)autonomously.[18] Robots are found in many settings, from factories to homes. They can be machine-like, animal-like, or human-like. Progress in robotics has lagged technologies like AI, due to the difficulties of creating machines that can sense and appropriately react in complex real-world environments. However, robotics is accelerating due to advances in computer science.

*Example of an issue in this area:* What kinds of moral 'reasoning' abilities or rules, if any, should autonomous robots be given when they interact with us?

This is, by no means, an exhaustive list. Also, we should be mindful that these subfields tend to overlap with one another based on similar features and a common evolution. Compare, for example, how a *tomato* is treated as a vegetable when it comes to ordering salad and as a fruit when it comes to botany. In consequence, discussion about whether a decision-support tool for management judgements is fair and equitable can draw on fields AI, ML, HCI, and IS.

A natural question here is, who is responsible for what, when it comes to potential ethical pitfalls and concerns in the products of computing – from software to robots? Also, given these disciplinary differences (and similarities), what can practitioners, technologists, and academics agree on, given the 'fuzzy' nature of the ever-expanding discipline?

## What Can We Agree On? The Technologists' View

From a technology practitioner's standpoint – whether we are programmers, academics, or engineers – we can surely all agree that tech has its flaws. We may also agree that *no single person* can fix the myriads of flaws in tech, and that it is not the exclusive domain of any specific discipline or speciality in computing.

If we are lucky, we are made aware of such issues *before* they arise – say, when someone decides to pause the deployment of a potentially problematic system prior to it being rolled out into everyday use. Often, however, problems are noticed *after* the fact; and quite often, these problems enter the public consciousness thanks to press reports or courageous whistleblowers.

To orient our thinking, let's consider a concrete, current, issue in technology and digital ethics. In 2021, the *Wall Street Journal* published an exposé on Facebook (now Meta) – *The Facebook Files* – which reported several ethical issues plaguing the platform. Later, it was revealed that a Facebook employee named Frances Haugen was the whistleblower who brought this issue to light.

To summarise one of the issues raised from Haugen's leak, as reported in The Facebook Files:

> Facebook allowed teenagers to be exposed to content about unrealistic body image, which affected their mental health and caused them to be vulnerable.[19]

From the technology side, one can argue that several subfields had a role in how this came to be. For starters, practitioners from *AI* and *ML* played a role in the design, development, and deployment of the recommender systems (such as a news feed page). In this instance, Facebook's recommendation of unrealistic body images to vulnerable users – in other words, their misguided *optimisation of content* – were and are clearly harmful ones. One could also argue that *CV* plays a role here, as the social media platform could identify potentially problematic content (such as anorexia-related images) with CV and then make meaningful interventions.

*HCI* research could also come into play, including by enabling interventions such as slowing down (or 'adding friction' to) user interaction, thereby reducing teenagers' rate of consumption of content. An example would be disabling the 'infinite scroll' or 'autoplay' features on major social platforms. Last but not least, social media platforms are inherently 'information systems' that are used by communities and societies; therefore, *IS* research on use/adoption of, and addiction to, technology can help regulators and parents know how these technologies gain traction amongst younger users.

Of course, it takes the concerted effort of not just technologists, but also policymakers, regulators, and other stakeholders, to solve such multi-faceted problems. Furthermore, computing sciences can only do so much in identifying, understanding, and tackling techno-ethical problems without the help of disciplines such as law, sociology, cultural studies, anthropology, psychology, and philosophy – to name just a few.

We will revisit this point – particularly, what sort of concerted effort is required to ameliorate current technological harms – towards the end of the book (Chapter 7). But, for now, we aim to equip you with the skills you need to understand such (digital) ethical dilemmas or problems. For this, we need to unpack the toolkit of the philosopher.

## How Is Digital Ethics Related to Philosophy?

### On Ethics

This book is about moving from philosophy to practice. We therefore should know something about philosophical ideas. In particular, we need to hear from moral philosophy, or, as it is often simply called, 'ethics'.[20] Ethics, incidentally, can also refer to societal or individual morals or morality – to our beliefs about right and wrong behaviour and about good and bad character. We can thus use the word 'ethics' to refer either to moral philosophy or to people's morals or morality.

In this book, will be discussing ideas from moral philosophy. Having some understanding of several major philosophical ideas and frameworks will allow us to more fully understand digital technologies. With these conceptual tools, we can better argue about and analyse ethical questions that digital technologies create.

Ethics or moral philosophy is a big subject, so we can only scratch the surface in this book. Indeed, for millennia, philosophers have been discussing questions about how people should behave, how we ought to act, and what sort of characters we should aspire to have. Instead of going into great technical detail about ethics or moral philosophy, then, our more modest aim will be to provide an overview – one that will be practically useful for thinking more reflectively about issues in digital ethics.

To be clear up front, in this book, we will only be concentrating on one type of ethics, namely Western moral philosophy. Western philosophy began over 2500 years ago with the ancient Greeks, especially with the major Greek figures of Socrates, Plato, and Aristotle.[21] Since then, Western philosophy has flourished in Europe, the UK, the US, Australia, and other countries up to the present day.

Focusing on western philosophy is obviously a limitation in our approach, since there are many and varied insights to be found in other traditions, like eastern and indigenous philosophies.[22] We should also point out that historically, the western tradition mostly marginalised non-male thinkers; the most well-knows names in that tradition are those of White men: such as Augustine, Aquinas, Hume, Kant, Hegel, Nietzsche, Marx, Wittgenstein, and others. Women philosophers are fortunately now becoming more recognised and appreciated.[23]

Nonetheless, we will focus on the Western philosophical tradition partly because that is the tradition we are most familiar with and partly because it features heavily in contemporary debate about digital technology, including in leading journals, books, and conferences related to digital ethics.

## On Philosophy

To give now some further perspective of a different kind, we can also point out that ethics is only one part of philosophy more generally. Philosophy itself has several other main branches. These include the branches of aesthetics, political philosophy, metaphysics, philosophy of mind, logic, and epistemology. Some, if not all, of these areas of inquiry are relevant to digital ethics. For example, political philosophy is now being applied to questions about the broader social implications of digital technologies. This includes questions about, for example, how the World Wide Web, social media, and the metaverse will, should, or ought not to affect our community and its various institutions, such as democracy and freedom of speech.

Political philosophy will certainly have something to say, for example, about the explosion of the many, many voices on the internet since the late 1990s and the creation of 'filter bubbles', toxic content, radicalisation, and misinformation.

Internet founders such as Tim Berners-Lee and early internet adopters and proponents expected or hoped that the web would produce greater opportunities for ordinary, less powerful people to be respectfully listened to. Some people expected or hoped for a more informed population, stronger democracies, and the fall of authoritarian regimes. It has not exactly gone that way. Political philosophers are asking whether and how the evolution of the World Wide Web, and the various platforms and billionaire tech titans it gave rise to, can satisfactorily coexist with certain vital political values and practices, such as liberty of expression, protection from harm, discrimination, and hatred, and a well-informed voting citizenry.

Philosophy of mind, another branch of the philosophy tree, is also pertinent to some questions in digital ethics. For example, some scholars think that we might soon have moral duties to AI or robots if they become aware or conscious.[24] If machines do become truly conscious, these philosophers think, we might have duties to avoid switching them off, scrambling their circuits, or preventing them from cloning their programs. Imagine that these intelligent machines produce text or speech begging us not to harm them while also giving reasons why flicking off the switch, or stopping them from reproducing, would be a moral travesty or crime. Discriminating against such artificial intelligences, some will believe, could be a new form of human prejudice, similar to racism or sexism.[25]

But could such machines really be aware or conscious? The concept of consciousness, it turns out, is a very difficult one and there is much disagreement about its conditions. For example, some philosophers argue that true (if minimal) consciousness requires the ability to *feel* – not just to, say, intelligently solve tricky cognitive problems, as some generative AI now can. Some philosophers would go on to say that being able to have phenomenal feelings or awareness requires a biological body, or at least a biological brain, not merely, say, a 'positronic' brain (to use sci-fi writer Isaac Asimov's term) composed of circuits or other inorganic materials.

Others think, on the contrary, that conscious feeling and awareness is compatible with being a nonbiological entity made from silicon, metal, and plastic, in the right sort of configuration and with the right sort of functional properties. We might imagine here a humanoid robot fitted with advanced AI that, as in some science fiction stories, engages us in conversation about what it is like to feel the sun on its soft plastic skin, or what the experience is like of plugging into a power source and receiving an instant knowledge update.

## Great Minds Think Alike? (or Not?)

Despite the relevance of these – and other – branches of philosophy to digital ethics, it is moral philosophy that preoccupies us in this book. We have a few other points to make to give us additional perspective on ethics. Moral philosophy, like philosophy generally, is a controversial subject. That is, philosophers vigorously disagree about, for example, which is the best moral theory – or systemic framework for guiding our behaviour – of the several theories on offer. They even disagree on how to go about deciding and arguing which moral perspective or framework is the best or most defensible.

This makes moral philosophy (and philosophy generally) more controversial than most other disciplines. After all, you don't usually see physicists arguing about the validity of the scientific method, or mathematicians arguing about whether algebra is, or is not, nonsense. (Which is not to say that there aren't other types of vigorous disagreement in physics or mathematics.)

Yet some philosophers think, for example, that the manner in which other philosophers 'do' ethics is in some ways fundamentally misguided. If you've ever heard philosophers discuss matters from their discipline, you might know that they often like to disagree with each other about seemingly anything and everything! This makes philosophy exciting for some people and somewhat frustrating for others.

Yet despite the prevalence of disagreement about many matters in philosophy and ethics, most philosophers would say that some moral perspectives are better than others. And they are better, most philosophers believe, for reasons which people can outline and explain to one another. Most moral philosophers would claim that it is possible to coherently discuss, carefully offer arguments, and debate at length about moral positions and situations. In thinking and arguing about ethics, we can, they hold, achieve a greater understanding of morality and our moral lives. This includes our lives of increasing enmeshment with digital technologies.

## How Does One 'Do' Ethics?

Having provided a gentle segue into the topic of ethics, let's dive right in, to give you a taste of our exploration in the next few chapters.

We will now reconsider our example of Facebook, whose teenage users were affected by content about unrealistic body image that was found in their customised feeds.

How would a philosopher, or applied ethicist, approach this controversy? They might first say that a **moral agent** – simply put, a human such as you or me, who is able to differentiate between *right* and *wrong* – could believe that:

> Facebook has done the wrong thing to some teenagers through the design of their technology.

Before worrying about whether this belief is justified, we will need to take one quick step back, and judge whether ethics and morality can themselves be justified!

Cue **metaethics**: The study of the meaning of moral ideas and the status of moral beliefs.[26] Meta-ethicists investigate whether moral beliefs about right and wrong action can really be true or false, justified or mistaken. You might have heard the sceptical view (or you might even have had the thought yourself) that moral beliefs cannot really be correct or incorrect because they are merely a matter of opinion or feeling. It is worth explaining this view, in case it prevents us from taking moral argument seriously.

## Ethical Subjectivism vs Relativism

In metaethics, this first view, which we can label **ethical subjectivism** – sometimes called 'emotivism' – insists that moral judgements are not objective and have no cognitive basis.[27] Unlike, say, judgements about the physical world or judgements within mathematics, moral judgements cannot get things right or wrong. They are simply expressions of feeling or emotion. According to this kind of ethical subjectivism, *a moral belief is not correct or incorrect*. It is neither mistaken nor the truth. It is more-or-less simply a negative reaction or feeling about, for example, Facebook's behaviour. Moral judgements are more like mere feelings about, say, your favourite colour or favourite type of junk food.

This subjectivist view would mean that critics of Facebook are not more, and not less, close to the truth than defenders of Facebook. For ethical subjectivists, moral argument overwhelmingly comes down to clashes of emotional or affective reaction. While some people warm to this idea, many feel that there must be more to morality than brute expressions of feeling, even if feeling and emotion turn out to be important or even essential aspects of ethics.

There is a different sceptical view to ethical subjectivism, which is also attractive to some people. This second view is called **ethical relativism**.[28] According to a common version of this position, moral judgements about what is good or bad can indeed be correct or incorrect, but the correctness or incorrectness of those judgements lies entirely in what a majority of people in society think is morally good or bad.

For example, if most people in society think Facebook did no wrong in allowing their algorithms to push content about unrealistic body image to teens, then it would be correct to say that Facebook was beyond reproach.

Conversely, if society judges that Facebook's behaviour was unethical, then the judgement that Facebook did wrong is thus correct (and the judgement it did not do wrong is incorrect).

This culturally relativistic way of understanding ethics or morality has the result that if society changes its mind, then the moral truth necessarily changes. Suppose society holds at one time that that Facebook did no wrong, but shortly afterwards – e.g., after Mark Zuckerberg apologised in the US Senate to the parents of children harmed by Facebook – society then holds that what Facebook did was terribly wrong.

In this situation, ethical relativism would imply that what Facebook did was *both not wrong and wrong*! Moral correctness or truth simply depends on what people say and on which group of people or time period we are speaking about. The fact that the very same behaviour, with the exact same consequences and features, can be deemed both wrong *and* not wrong (or right) is a very strange implication of this metaethical position. Moreover, we don't have to accept it.

## Navigating Tensions

Here is an important observation: Most of us would agree that personal feelings and the views of the public can sometimes be mistaken. In the course of history, society and its individuals have at times been ethically misguided, even though the views they held once seemed perfectly obvious and natural to them. We need only think here of earlier views that we now regard as sexist, racist, homophobic, ableist, ageist, and so on.

Likewise, what is morally right can also sometimes conflict with what the law says. In many places, the law once supported (or in some cases still supports) various kinds of moral prejudices and immoral practices, such as slavery and the oppression of women and minorities. Law has also sometimes forbidden behaviours that we now regard as morally vital, like freedom to criticise the government or to practice our own religion or core beliefs. Because law and the state are so powerful, they can shape our own and society's moral viewpoints. This shaping can occur in good ways when the law is morally right, and in bad ways when the law is morally wrong. Thus, despite the difference between ethics and law, they are related to each other.

Ethical reflection may sometimes suggest that laws offering more protection to individuals or groups need to be introduced. The EU, for example, thought this necessary when it introduced its General Data Protection Regulation law (GDPR),[29] which aimed to give citizens more privacy and control over their digital data. Contrariwise, ethical appraisal may suggest that some existing laws are sufficiently unfair, harmful, counterproductive, or freedom-limiting that they should be modified or repealed.

In sum, what is morally right can sometimes clash, and sometimes chime, with what is legal, customary, socially accepted, and seemingly natural. And what is morally right can conflict or align with our initial emotional reactions. To justify our moral positions, especially when controversy and disagreement arises, we need to go beyond reflex judgements and gut feelings, and beyond prevailing social attitudes, to produce reasons and arguments to back up a point of view. For the above sorts of reasons, ethical subjectivism and ethical relativism are on shaky ground. This is where normative ethics comes in.

### Normative Ethics

Unlike metaethics, **normative ethics**[30] is less concerned with the nature of moral judgements – such as whether they can be valid or are merely subjective or relative to society – and more concerned about examining which moral judgements, positions, or perspectives, are the right or best ones. Thus, within normative ethics we find discussion about so-called ethical or moral theories,[31] which we will dive into shortly. As we will see, these theories highlight central ideas in ethics such as the consequences of our actions, basic duties and obligations, good character traits and dispositions, and tendencies to care about fellow human beings.

We can also find in normative ethics a number of fields of practical or applied ethics. These more practically orientated branches of moral philosophy include business ethics, environmental ethics, bioethics, animal ethics, professional ethics, and legal ethics. Each of these applied fields can intersect and overlap with digital ethics. For example, medical AI raises ethical questions for bioethics, such as whether doctors should use diagnostic AI that might be inaccurate or potentially biased against some patient groups.

To give another example, big data centres and the training of large language models raise questions for environmental ethics, about technology's increasing impacts on ecosystems and climate, and the sustainable use of resources such as minerals and water.[32] Bioethics and environmental ethics require both knowledge of moral philosophy and, very often, knowledge about medical science and environmental science, respectively.

Similarly, digital ethics requires both knowledge about ethics and knowledge about the nature and workings of technology.[33] For example, knowing enough about ML is important for understanding problems of opaqueness and bias in AI algorithms. Digital ethics can also require some understanding of socio-technical systems – that is, how technology and society relate to and affect each other.

## What Can We Agree On: The Philosophers' View

Earlier, we observed that philosophers disagree on many things. Disagreement on fundamental matters, including ethics, is part of philosophy itself. Does this mean that it is hopeless to think that philosophy can help us with ethical thinking? Actually, it doesn't mean that.

To return momentarily to metaethics, we can note that most moral philosophers are not fans of *either* ethical subjectivism or ethical relativism.

Philosophers generally believe that an individual's moral feelings or views (in the case of ethical subjectivism) can be mistaken or correct, and that society's moral opinion or views (in the case of ethical relativism) are *insufficient* to determine the correctness or truth of a moral view.

A key reason for this, as we also noted a few paragraphs back, is that we seem to be able to give reasons and arguments for moral views and positions. These reasons and argument go beyond brute feelings or social attitudes. Thus, what determines whether a moral judgement is correct is, at least in large part, the strength of the *reasons* which can be advanced for them.

## Providing Justifications for Moral Positions

Let's return to our Facebook example to illustrate. Someone might offer the following reasons to justify their position that Facebook did indeed do the morally wrong thing in deploying its algorithm:

1. In pushing content about body image that made some teenagers feel ashamed or inadequate, Facebook caused a great deal of long-lasting psychological harm to many human beings. Many of these young people were at an emotionally fragile stage of life.
2. While these young users benefitted in some ways from social media, they were also seriously harmed by it.
3. Furthermore, Facebook knew that these major harms could occur and yet went ahead anyway to advance their business interests.
4. True, pursuing profit may not be wrong in itself and may be *good* in various ways for society. It creates employment for example, which is vital to our lives.
5. But commercial profit should not be pursued regardless of the consequences to others. In fact, causing or allowing significant harm to befall innocent human beings merely for profit goes to the heart of what ethics is all about.
6. Facebook's behaviour and those of its executives also shows a deficiency of compassion and justice and the unfortunate presence of significant greed and callousness.
7. In cleverly pushing harmful content without informing young users, the social media platform wrongly used some of its customers without their agreement or consent.
8. The company failed in its duty of care to individuals who, by using the platform, were also actually benefitting that social media company.

These are just some possible reasons that someone who is engaging in ethical debate about the Facebook incident could put forward. And, as we shall soon see, some elements of the reasons just given have been analysed, clarified, and put more systematically by philosophers, usually in the form of 'moral theories'.

This list of reasons at least suggests that determining the validity or correctness of moral judgements goes beyond observing outbursts of personal feeling, and beyond taking polls of public sentiment to gauge what most people think. In fact, it is important to see that the morally right point of view can sometimes conflict with what many or most people hold, and with our own initial emotional reactions as well.

So far, we have given you a glimpse of philosophy and how we might begin to go about following a reasoned approach to ethical questions. In the next chapter, we will briefly press the pause button on philosophy to confront a list of issues that scholars and practitioners agree are appearing in our contemporary digital landscape.

## Notes

1. From the Oxford Dictionary of English, as reproduced in the Dictionary app in MacOS.
2. A sad example is the Tuskegee Syphilis Study (1930s to 1970s), which had an array of unacceptable practices including not having "informed consent from participants" and withholding treatment "even after it was easily available". This will be briefly discussed in the context of Chapter 5. For more information, see, e.g., the US CDC writeup at https://www.cdc.gov/tuskegee/about/index.html.
3. See Helen Rosner's article in The New Yorker: Rosner (2021, July 17). An ethical analysis can be found in Cheong (2023)
4. Chrisman et al. (2013).
5. Cheong, Leins, & Coghlan (2021).
6. In reality, there are multiple definitions. A good reference is the authoritative AI textbook: Russell & Norvig (2010).
7. Based on the latest definition by the OECD. See OECD.AI. (2023, November 29).
8. This ensemble definition is based mainly on McKinsey & Company (2024, April 2). The OECD has a similar definition which is close to the one provided by McKinsey: OECD (n.d.).
9. Aho & Ullman (1994).
10. From the Oxford Dictionary of English, as reproduced in the Dictionary app in MacOS.
11. This ensemble definition of CV is by Microsoft and IBM, respectively: Microsoft Azure (n.d.) and IBM (2021, July 27).
12. This ensemble definition of HCI comes from one of the authors' experiences as an HCI lecturer for approximately half a decade.
13. Rainer & Prince (2021).
14. Burkov (2019).
15. Adapted from definitions by Oracle and IBM: Eppright (2021, March 25) and Stryker & Holdsworth (2021, September 23).

16. IEEE (2022).
17. Sommerville (2015).
18. ISO (n.d.).
19. For the Wall Street Journal story, see Wall Street Journal (2021, October 1). For more on the topic, we recommend reading Haugen's memoir: Haugen (2023).
20. Wolff (2020).
21. Pojman (1995).
22. Harrison (2019).
23. Waith (1995).
24. On the debate about robots, rights, and moral duties, see: Bryson (2010), Gunkel (2024), and Coeckelbergh (2010).
25. Sebo (2025).
26. Fisher (2014).
27. Rachels & Rachels (2012).
28. Levy (2014).
29. The GDPR is more formally known as: Regulation (EU) 2016/679 of the European Parliament and of the Council of 27 April 2016 on the Protection of Natural Persons with Regard to the Processing of Personal Data and on the Free Movement of Such Data, and Repealing Directive 95/46/EC (General Data Protection Regulation) (Text with EEA Relevance), 119 OJ L (2016).
30. Kagan (1998).
31. Gensler (2017).
32. Crawford (2021). See also Chapter 3.
33. Beever, McDaniel, & Stanlick (2019).

## References


Aho, A. V., & Ullman, J. D. (1994). Foundations of Computer Science. Computer Science Press.

Beever, J., McDaniel, R., & Stanlick, N. A. (2019). Understanding Digital Ethics: Cases and Contexts. Routledge.

Bryson, J. J. (2010). Robots should be slaves. In Y. Wilks (Ed.), Close Engagements with Artificial Companions: Key Social, Psychological, Ethical and Design Issues (pp. 63–74). John Benjamins Publishing.

Burkov, A. (2019). The Hundred-page Machine Learning Book. Andriy Burkov.

Cheong, M. (2023). Chef of the future? In S. Calef (Ed.), Anthony Bourdain and Philosophy (pp. 191–200). Carus Books.

Cheong, M., Leins, K., & Coghlan, S. (2021). Computer science communities: Who is speaking, and who is listening to the women? Using an ethics of care to promote diverse voices. Proceedings of the 2021 ACM Conference on Fairness, Accountability, and Transparency (pp. 106–115). https://doi.org/10.1145/3442188.3445874

Chrisman, M., Pritchard, D., Fletcher, G., Mason, E., Lavelle, J. S., Massimi, M., Richmond, A., & Ward, D. (2013). Philosophy for Everyone(p. viii). Routledge.

Coeckelbergh, M. (2010). Robot rights? Towards a social-relational justification of moral consideration. Ethics and Information Technology, 12(3), 209–221. https://doi.org/10.1007/s10676-010-9235-5

Crawford, K. (2021). The Atlas of AI: Power, Politics, and the Planetary Costs of Artificial Intelligence. Yale University Press.
Eppright, C. (2021, March 25). What is Natural Language Processing (NLP)? Oracle. https://www.oracle.com/au/artificial-intelligence/what-is-natural-language-processing/
Fisher, A. (2014). Metaethics: An Introduction. Routledge.
Gensler, H. J. (2017). Ethics: A Contemporary Introduction. Routledge.
Gunkel, D. J. (2024). Robot Rights. MIT Press. https://mitpress.mit.edu/9780262551571/robot-rights/
Harrison, V. (2019). Eastern Philosophy: The Basics (2nd ed.). Routledge.
Haugen, F. (2023). The Power of One: How I Found the Strength to Tell the Truth and Why I Blew the Whistle on Facebook (1st ed.). Little, Brown and Company.
IBM. (2021, July 27). What is Computer Vision? IBM. https://www.ibm.com/think/topics/computer-vision
IEEE. (2022). IEEE/ISO/IEC International Standard for Software, systems and enterprise—Architecture Description. IEEE Standards Association. https://standards.ieee.org/ieee/42010/6846/
ISO. (n.d.). ISO 8373:2021(en), Robotics—Vocabulary. https://www.iso.org/obp/ui/en/#iso:std:iso:8373:ed-3:v1:en
Kagan, S. (1998). Normative Ethics. Routledge.
Levy, N. (2014). Moral Relativism: A Short Introduction. Oneworld Publications.
McKinsey & Company. (2024, April 2). What is ChatGPT, DALL-E, and Generative AI? https://www.mckinsey.com/featured-insights/mckinsey-explainers/what-is-generative-ai
Microsoft Azure. (n.d.). What is Computer Vision? Retrieved March 17, 2025, from https://azure.microsoft.com/en-au/resources/cloud-computing-dictionary/what-is-computer-vision
OECD. (n.d.) Generative AI. https://www.oecd.org/en/topics/sub-issues/generative-ai.html
OECD.AI. (2023, November 29). Updates to the OECD's definition of an AI system explained. OECD.AI. https://oecd.ai/en/wonk/ai-system-definition-update.
Pojman, L. P. (1995). Ethical Theory: Classical and Contemporary Readings. Wadsworth Publishing Co. Inc.
Rachels, J., & Rachels, S. (2012). The Elements of Moral Philosophy (7th ed.). McGraw Hill.
Rainer, R. K., & Prince, B. (2021). Introduction to Information Systems. John Wiley & Sons.
Regulation (EU) 2016/679 of the European Parliament and of the Council of 27 April 2016 on the Protection of Natural Persons with Regard to the Processing of Personal Data and on the Free Movement of Such Data, and Repealing Directive 95/46/EC (General Data Protection Regulation) (Text with EEA Relevance), 119 OJ L (2016). http://data.europa.eu/eli/reg/2016/679/oj/eng
Rosner, H. (2021, July 17). The Ethics of a Deepfake Anthony Bourdain Voice. The New Yorker. https://www.newyorker.com/culture/annals-of-gastronomy/the-ethics-of-a-deepfake-anthony-bourdain-voice.
Russell, S. J., & Norvig, P. (2010). Artificial Intelligence: A Modern Approach (3rd ed.). Prentice Hall.
Sebo, J. (2025). The Moral Circle: Who Matters, What Matters, and Why (a Norton Short) (1st ed). W. W. Norton & Company, Inc.

Sommerville, I. (2015). Software Engineering (10th ed.). Pearson Education.
Stryker, C., & Holdsworth, J. (2021, September 23). What is NLP (Natural Language Processing)? IBM. https://www.ibm.com/think/topics/natural-language-processing
Waith, M. E. (1995). A History of Women Philosophers: Contemporary Women Philosophers, 1900-Today. Springer Science & Business Media.
Wall Street Journal. (2021, October 1). The Facebook Files. Wall Street Journal. https://www.wsj.com/articles/the-facebook-files-11631713039
Wolff, J. (2020). An Introduction to Moral Philosophy (2nd ed.). W. W. Norton & Company.

# 2

# *The Changing State of Play: What Are We Concerned With?*

The conversation about digital technologies sometimes revolves around the *good* versus the *bad*. Sometimes it concerns the ways in which existing long-held practices are *changed* by technology or even rendered *obsolete*. And sometimes discussion relates to why some technologies shouldn't or couldn't last the test of time.

Consider how much we rely on tech nowadays – say, when our robot vacuum sends notifications to our smartwatch, when our car sends our driving patterns back to the manufacturer for diagnostic purposes, or when we make a video call on our mobile phones. In other words, our dependence on digital technologies is *pervasive, ubiquitous,* or simply *in the everyday*.

## Ubiquitous Computing

Digital devices, infrastructure, and software work together – in essence, performing the act of *computing* – in our daily lives.

In fact, there is a field of study called ***ubiquitous computing* (*ubicomp*) or *pervasive computing*** that falls across many computing subfields. The term 'ubiquitous computing' was coined by Mark Weiser in the late 1980s. Weiser could have well predicted the current state of computing and digital devices when he wrote his *Scientific American* article, '*The Computer for the 21st Century*' in 1991. There he proposed …

> a new way of thinking about computers in the world, one that takes into account the natural human environment and allows the computers themselves to vanish into the background.[1]

Think about this for a second. When we are considering the examples above (from robot vacuums, to car diagnostics, to video calling), we usually don't remember that there is a series of interlinked computers involved in the processes: from miniaturised sensors, to cloud computing servers, to our mobile phones providing the interface to apps. (We recommend checking out


DOI: 10.1201/9781003393511-2

the highly detailed visualisation by Kate Crawford and Vladan Joler in mapping out the *'Anatomy of an AI System'*.[2])

When defining the term 'ubiquitous computing' in 1988, Weiser, in effect, described what we have in this example: "It is invisible, everywhere computing that does not live on a personal device of any sort, but is in the woodwork everywhere".[3]

As these technologies are everywhere – beyond the box or laptop we traditionally deem as a 'computer' – naturally, we will be using or interacting with them quite often. This ubiquity, and hiddenness, makes ethics a foreground concern.

## Big Data

It's not just the fact that computers are everywhere nowadays: The amount of valuable data we consume, remix, and produce is growing extremely fast.

This is where we can draw inspiration from studies in ***big data***. Big data can be characterised by the so-called '3 Vs' – ***volume, velocity****, and* ***variety*** – proposed by Doug Laney of META[4] Group (now Gartner) as early as 2001.[5]

*Volume* refers to the increasing amount of data generated over time. Data is no longer limited to, say, that deliberately generated by a user in response to an information need (such as a question or prompt). This is a far cry from early days of manual record keeping. For example, completing a form for a doctor's appointment never captured where you were sitting or how fast your pen moved across the paper. By comparison, even simply clicking your mouse or tapping the screen can generate data (telemetry) collected by various apps. And, to take another example, sensors and CCTV cameras that hoover data are ubiquitous nowadays.

*Variety* refers to the wide range of data sources that capture data about us. What used to be just a doctor's form, for instance, now extends to the telemetry data captured by the web application provider and security surveillance hardware. The variety of data about oneself becomes even more concerning when these different sources or strands of data are combined to provide a more detailed picture of our behaviour and attitudes.

Finally, *velocity* refers to the speed at which data is generated, processed, and stored. The concern here goes beyond data volume and variety. Data can now be reconfigured and spread across the internet at speed. Consider how digital deepfakes no longer require high production costs or time-consuming distribution, but can instead be rapidly created and transmitted around the world on social media.

The 3V's make digital ethics more pertinent than ever: Happenings in the digital space can go wrong fast (velocity), in more ways than one (variety), and repeatedly (volume); and this potential is ever increasing.

## Reeling in the Years[6]

Given the pervasiveness of technology and the 3Vs of data, how do we even start to describe what's wrong with technology? How in a short primer do we make these multifarious issues accessible for the reader?

As educators in digital ethics, an adage we like to employ is that *the history of digital technology is a subset of the history of culture and of society*. By witnessing how technology has changed through the years, we can contextualise and better understand relevant sociocultural and ethical issues.

Though not exhaustive, the following historical overview illustrates the scope and depth of the problem we tackle in this book. In these examples, we touch on the previously mentioned subfields in computing (Chapter 1).[7]

### 1940s: Dawn of Modern Computing

In the 1940s, the term 'robot' entered the cultural zeitgeist with Isaac Asimov's publication of the *three laws of robotics*.[8] The term 'robot' originated from author Karel Čapek's 1920's play *R.U.R. (Rossum's Universal Robots)*.[9] Obviously, back then, there were no robots sufficiently advanced to vacuum our house or serve food at a restaurant. But Asimov's science-fiction laws about robots being 'programmed' not to harm humans, to follow our commands, and to protect their own existence (in that order) were an early version of a philosophy governing robot–human interaction.

In 1950, the great computer pioneer Alan Turing published his philosophical idea of an *'imitation game'*. In this game, a machine is tested to see if it can fool people who ask it questions (but who cannot see it) that it is human, not machine. The so-called *'Turing Test'* is supposed to show whether a computing machine can be said to 'think'.[10]

### 1950s: Culture and the Beginning of AI

Before the introduction of advanced automated computers such as the ENIAC[11] in 1945, *computers were actually human jobs*. These jobs were staffed by women, particularly African-American women.[12] Clive Thompson of the *Smithsonian Magazine* writes:

> Women were welcome as computers partly because the work was viewed as a dull, low-status activity. Men with elite educations generally wanted no part in it.[13]

The *Dartmouth Summer Research Project on Artificial Intelligence* in 1956 coined the term 'Artificial Intelligence (AI)'. It was organised by pioneers – all White men – such as John McCarthy, Marvin Minsky, Nathaniel Rochester, and Claude Shannon. Although scientifically focused, this conference raised the philosophical issue of whether computers can replicate human thought.[14]

Around this time, Norbert Weiner, a pioneer in the field of cybernetics, explored the potential of automation to change society. His book *The Human Use of Human Beings* contemplated risks that automation might pose to society (including loss of control or overdependence, issues we'll revisit in Chapter 7).[15]

## 1960s: Promises and Perils of Tech

In the following decade, Joseph Weizenbaum invented the early chatbot *ELIZA* using basic computerised rules. However, one ethical issue he raised is the human ascription of 'humanness' (*anthropomorphism*) to the chatbot, to the point of people using it for therapy and emotional connection. Anecdotally, Weizenbaum's own secretary told him to "leave the room while she finished her consultation".[16] Ironically, ELIZA was merely based on simple programming logic and could not reason.[17]

The focus on technology ethics in the 1960s was more on the advancements in *engineering*, particularly in the sphere of *weaponry* in the postwar period.[18]

## 1970s: Hints of Pessimism

Notwithstanding early excitement about AI in the 1956 Dartmouth project, the 1970s brought a sobering realisation. Sir James Lighthill's report, *Artificial Intelligence: A General Survey*, presented a pessimistic view of AI's progress. The *Lighthill Report* damningly said:

> Most workers in AI research and in related fields confess to a pronounced feeling of disappointment in what has been achieved in the past twenty-five years. Workers entered the field around 1950, and even around 1960, with high hopes that are very far from having been realised in 1972. In no part of the field have the discoveries made so far produced the major impact that was then promised.[19]

This would mark another period of stagnation in AI research, known as an *AI winter*. AI's great progress in the late 2010s and 2020s is the latest *AI summer*. As the seasonal analogy implies, these winters and summers come and go.

Digital interfaces – the object of study of fields such as *HCI* – came under scrutiny after seminal incidents like the *Three Mile Island* nuclear accident in 1979.[20] A contributing factor to this partial nuclear meltdown was the user-facing instruments and alarms (i.e., the 'user interface'), which offered conflicting and confusing information to operators, without offering them remedies.[21]

## 1980s: Ethics and Dangerous Tech

Around the 1980s, researchers started taking seriously ethical concerns[22] regarding the increasing use and deployment of technology. Examples include Richard Mason's (1986) identification of four ethical concerns, or *PAPA* (*Privacy, Accuracy, Property, and Accessibility*).[23] Similarly, B.J. Kotrewski and Charles Oppenheim (circa 1979) also highlighted concerns in their paper, *Ethics in information Science*, including "[accountable] use of work facilities … Confidentiality of information … Bias in information presented …".[24] These concerns are still relevant decades later.

The 1980s saw the (increasingly) interconnected nature of computers, including the use of storage devices and network connections, which enabled the transmission of computer *worms* and viruses. Illegal destruction, theft, or access to data or systems were also recognised as ethical issues.

The 1980s witnessed an infamous example of overconfidence in technology. *Therac-25* was a computer-controlled device that administered radiation therapy to cancer patients. This machine did not employ advanced AI techniques as machines were to do in the 2020s. Due to programming issues and a bad interface, the device incorrectly gave fatal radiation doses to some patients. Therac-25 is a textbook case in digital ethics[25]: Problems ranged from poorly managed safety checks and evaluations to what we now call automation bias.[26]

## 1990s: Resilience, Malignance, Surveillance

In the early 1990s, Linus Torvalds released the *Linux operating system*, now common in everyday devices. Torvalds's decided to **open source** his code, allowing people to access and freely use it without hefty fees. The open source ethos gives us an opportunity to make computing more affordable and bridge the ***digital divide***, especially in the Global South or Majority World.[27]

As computers supported more advanced technology, video games became more realistic, even before Windows 95 was launched. But this brought concerns about harms of video games and other multimedia content toward vulnerable users. A case in point is the video game DOOM, which was criticised and even banned in certain jurisdictions for promoting violence.[28]

As the Internet (Web 1.0) gained momentum, early browsers such as Netscape Navigator and Internet Explorer (which ended up having their *browser wars* for market share) were developed. With this came an interesting and enduring 'feature': online ads.[29] These ads became the focus of ethical and privacy concerns as platforms' began collecting copious data to perform user profiling and ad ***microtargeting***[30] (more in the next section).

The 1990s robot toy *Furby* was released to great fanfare due to its babbling and simulation of 'understanding' when children spoke to it. However, the issue of deploying such complicated (at the time) pieces of technology resulted in privacy and security concerns. The reports of surveillance, while anecdotal, [31]raised ongoing issues of privacy and consent for 'smart' or digital devices.

As the 1990s closed, some attention was given to the accessibility of the World Wide Web for all users. The *Web Content Accessibility Guidelines (WCAG)* were developed, offering web developers and maintainers a guide to make their websites more accessible, especially to those with disabilities.[32] Unfortunately, change was too slow. In Australia, Bruce Maguire, who was visually impaired, lodged a complaint against the Sydney Organising Committee for the Olympic Games (SOCOG). Maguire contended the Committee made it difficult for him to buy tickets by not offering accessible alternatives to their online option.[33] After an inadequate response, SOCOG eventually had to pay damages to Maguire. This landmark case increased ethical awareness about accessibility for websites.

## 2000s: New Tech in One's Pocket

In the early 2000s, the earliest mobile phones with built-in cameras were released. Though such devices would be ubiquitous in the 2010s and 2020s, this landmark innovation increased the availability of video (and audio) recording and transmission to practically anyone with a smartphone. The evolution of smartphones created issues of privacy, device addiction, and image-based abuse.

Cue the mid-2000s, with the introduction of early social networks or social media sites that variously combine text, videos, and images. While MySpace and Friendster did not survive across the decades, Facebook and YouTube certainly did. Facebook was originally meant for university students but became available to everyone.

Three related technological developments proved important. Firstly, Facebook introduced *News Feed*, based on a social recommender algorithm.[34] This feature heralded the hyperpersonalisation[35] of social media feeds. Secondly, *Pages* of celebrities, brands, and personalities – not just human users – could be followed or messaged on these platforms, giving them substantial social influence. Finally, Apple iPhone and the now-defunct BlackBerry set the scene for *smartphone-mediated, 'always on'* social media presence. Gone were the days of logging on from a bulky desktop.

Bitcoin emerged in the late 2000s. The world's first **cryptocurrency**[36] was associated with environmental impact (from 'mining' activities), illegal transactions, and scams. Also, around this time, Google around began research on its early versions of *self-driving car* technology.[37]

## 2010s: Social Media and AI Summers

Earlier social media platforms prioritised connecting with people you know. By contrast, Twitter (now X) and its alternatives – *microblogging platforms* – allowed people to broadly and rapidly broadcast their content to anyone

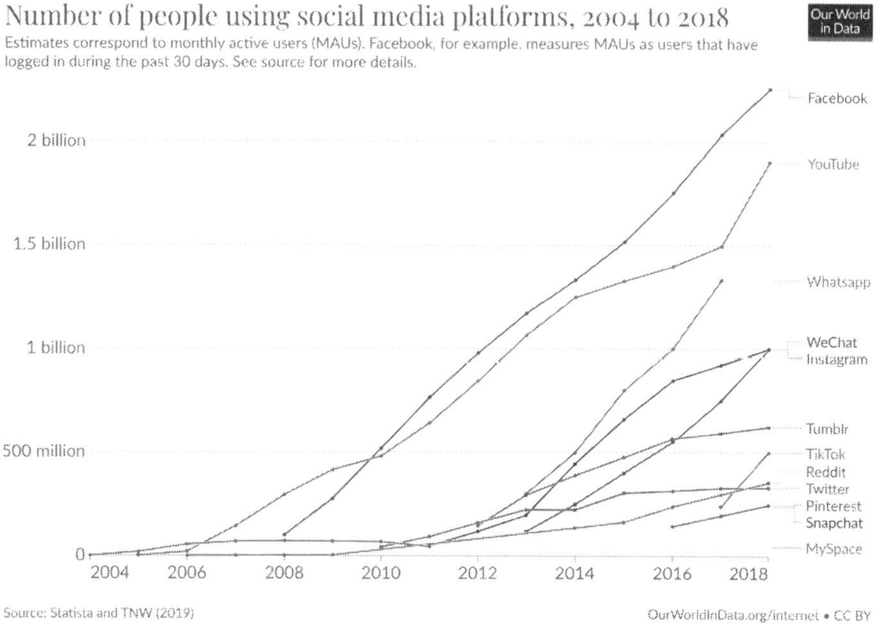


**FIGURE 2.1**
Graph by *Our World in Data*, comparing the growth trajectories of social media platforms such as Facebook, YouTube, and TikTok, from 2004 to 2018 inclusive. *(Image by Esteban Ortiz-Ospina (2019), based on sources from Statista and TNW: under the Creative Commons CC BY 4.0 – Attribution 4.0 International – license).*

interested in listening. The role of microblogging was prominent in the news in several *large-scale events* in the early 2010s, especially with Twitter. In the *Arab Spring* pro-democracy protests,[38] social media disseminated information and coordinated protest organisation. In the 2011 London Riots, social media both helped in inciting violence and cleaning up the damage.[39]

Social media use continues to grow. By 2013, YouTube and Facebook broke the 1 billion active user count per month (see Figure 2.1 for the trajectory of growth).[40]

By this stage, experts were noticing signs of a looming problem. Eli Pariser's 2011 book *The Filter Bubble: What the Internet is Hiding from You* introduced the term 'filter bubble' to describe what happens when platforms curate and personalise (the 'filter') content to yield a partial or lopsided worldview (the 'bubble').[41] When the user frequently engages with other users or content that aligns with a particular worldview (say, political affiliation), platforms show us related content to sustain our attention. This 'attention economy' effect results in users receiving less content that does not necessarily align with their worldviews, such as alternative political or medical opinion.

The 2010's also ushered in the present AI 'summer'. *Deep learning*[42] enjoyed a surge partly thanks to advancements in graphics card technology (hitherto used predominantly in video games). Some remarkable breakthroughs ensued. IBM's *Watson* supercomputer defeated human champions in the game show Jeopardy![43] Google's *AlphaGo* defeated the human champion Lee Sedol in the game of Go, to widespread astonishment.[44] The *transformer* architecture in deep learning proved to be the predecessor of powerful **Large Language Models (LLMs)** in the 2020s.

Unfortunately, such impressive advancements augured unwanted outcomes. The world was introduced to *deepfakes*[45] on the online forum Reddit – fake images, video, or audio created by AI, often to mimic real people – which were quickly used for altering faces in images, such as putting celebrities' faces onto actors in pornographic videos.[46]

With the advent of advanced automated classification and the increasing amount of data powering systems, unusual ethical issues were increasingly reported in the media. Consider some examples. Firstly, there was the potential for *gender bias* from AI-based recruitment tools. Amazon, which developed an experimental system to automate recruitment in a neutral manner, had to bin its resumé classifier for discriminating against women.[47] One reason is that such automated systems learn from patterns found in ingested data: the more men that were hired, the more the system would correlate, say, language typically used by men or resumé highlights such as 'boys' school', as measures of success.[48]

Secondly, there was inappropriate use of Facebook data in the *Cambridge Analytica scandal*. User data – and that of their contacts – were harvested without proper consent.[49] An estimated 30-80 million users were affected. The data was used for psychological profiling and 'microtargeting' of users on social platforms by political campaigns, in an attempt to influence elections.

Hearteningly, several initiatives in the 2010s raised public awareness of digital ethics and ethical initiatives. The European Union's *General Data Protection Regulation* (GDPR) was implemented in 2018, after being passed by the European Parliament in 2016.[50] According to the European Commission,

> The adoption of the GDPR was an essential step to strengthen individuals' fundamental rights in the digital age and facilitate business by clarifying rules for companies and public bodies in the digital single market.[51]

The GDPR even applies to organisations and companies outside the EU that collect the data of EU citizens. We encounter the GDPR in action when we encounter a cookie popup dialogue box, such as the one in Figure 2.2.

In academia, digital ethics movements began to take shape from the 2010s. The *Principles for Accountable Algorithms and a Social Impact Statement for Algorithms* was developed in 2016.[52] The popular *Fairness, Accountability, and Transparency in Machine Learning* (FAT/ML, now FAccT[53]) conference series became an annual event for researchers and practitioners.[54] Other conferences such as the *IEEE ETHICS* conference series[55] and the *AAAI/ACM Conference on AI, Ethics, and Society*[56] conference also started from 2014 onward.

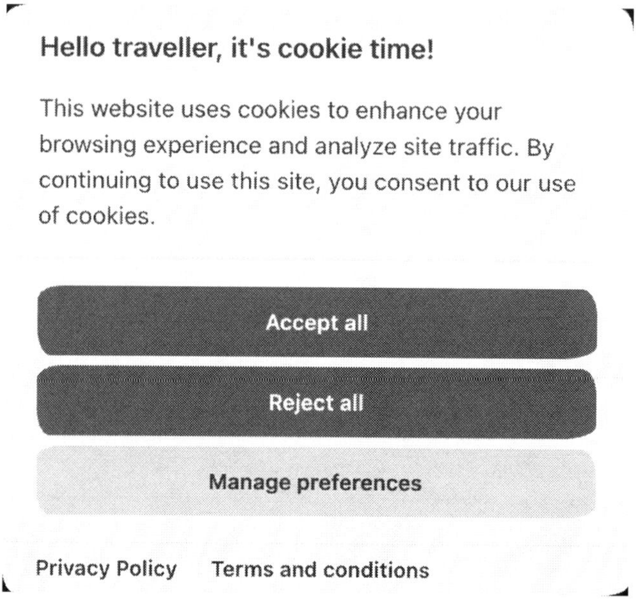


**FIGURE 2.2**
An example of a cookie popup prompt. (*Generated by the authors using the CookieConsent library developed by Orest Bida (2025) – https://github.com/orestbida/cookieconsent – software distributed under the MIT license).*

Best practice norms and initiatives for AI and digital technologies also emerged. These include the Asilomar Conference and *Asilomar AI Principles* (2017),[57] the IEEE's *Ethically Aligned Design: A Vision for Prioritizing Human Wellbeing with Autonomous and Intelligent Systems'* General Principles (2017),[58] and the *OECD AI Principles* (2019).[59] Increasingly, digital ethics principles were recognised as important (see Chapter 5).

## 2020s: COVID-19, ChatGPT, and Causes for Concern

The early 2020s was defined by change and uncertainty due to the COVID-19 pandemic. Physical distancing requirements meant that people increasingly turned to technology for work and leisure. The Internet became a vital source of information; but with this, came more-and-more misinformation and disinformation.[60]

Pandemic-related digital ethics issues[61] included:

- Usage of mobile devices for disease *contact tracing*, including concerns of privacy, data security, and autonomy, plus disadvantage to those who could not use mobile devices.
- The shift to a *contactless* economy (e.g., with credit card/mobile payments), with concerns about accessibility and equity for those

who could not afford a smartphone supporting mobile payments, or are unable to use a smartphone optimally.

- The general shift to working and studying remotely and questions about fairness and equity – *the digital divide* – since not everyone could afford enough digital devices, infrastructure, and connectivity for the household.
- *Working from home* and challenges of employee surveillance by organisations and videoconferencing technologies (e.g., preferences for turning one's webcam video off due to family commitments).
- The increase in the use of *online delivery* platforms and concerns about the welfare of gig workers as 'frontliners' during the pandemic.
- Students *studying online* who faced potential concerns to their privacy due to the use of technologies such as remote examination proctoring using webcams and AI exam cheating detection.
- COVID-19-related *untrustworthy information and online conspiracy theories* that hampered effective public health responses to the pandemic, including the dissemination of legitimate and scientific health advice.
- The use of *QR code location check-ins* and associated privacy concerns, especially since location data could easily be misused for surveillance or marketing.

Toward the end of the pandemic, the world was awed with the public release of OpenAI's ChatGPT at the close of 2022. This heralded the start of a wave of Generative AI or GenAI systems, including Google's Gemini, Anthropic's Claude, Mistral's Le Chat, DeepSeek, and Meta's Llama. These systems and their contemporaries could generate text, computer code, images, songs, videos, and more, given the right input prompts and other data files. The societal and ethical implications of these technologies are vast. A selection of such issues[62] include:

- Risk of generation of *harmful material*, such as extremist, racist, and illegal content.
- The use of GenAI systems for cheating and plagiarism (in education), creating *false claims of authorship*, or unauthorised *impersonation* of others.
- The *'pollution' of the Internet* with low-quality material.
- The potential harms of AI *outputs to vulnerable users*, including incitement to commit crimes or cause harm to themselves or others.[63]
- *Biases and harmful stereotypes* in the systems leading to discriminatory material and prejudices, especially when used in recommender systems.

- *Privacy and intellectual property concerns*, such as the use of confidential material as input, which might be used to improve future versions, or questionable ethics of using mass amounts of online data to train models.
- Broader societal and economic issues, such as *displacement of jobs*, harms of generative content toward *creative industries*, and risks of *loss of control due to overdependence on such systems*.

In response to this wave of worries, the Future of Life institute in 2023 called for a slowing down in the 'race' of producing more advanced Generative AI and other large-scale experiments in their *Pause Giant AI Experiments* open letter[64] (which we will revisit in Chapter 7).

Governments around the world are currently struggling to address the new upswell of AI technologies. Significantly, the *European Union AI Act* was brought into force in 2024.[65] Subsequently, the *AI Action Summit*[66] in 2025 showcased the *First International AI Safety Report*, written by "… 96 international AI experts who collaborated in an unprecedented effort to establish an internationally shared scientific understanding of risks from advanced AI and methods for managing them".[67]

* * *

This concludes our whirlwind tour of the history of digital technologies over 80-odd years and some key ethical issues that have accompanied it. Recent digital ethical issues have become increasingly varied and complicated. Nonetheless, ethical issues later in the timeline sometimes connect with prior ones. Consider: What does racism and sexism as far back as the 1950s have in common with AI-based hiring tools seen in the 2010s? An answer: Earlier bias/discrimination in society has entered into contemporary AI outputs.

Sometimes, though, earlier techno-ethical changes contrast with later ones. Consider: How does the 1990's Linux Open Source innovation *differ from* work-from-home initiatives during COVID-19? An answer: Each promised to increase equity, but while the Open Source movement helped close the digital divide, the pandemic work-from-home culture could increase the digital divide for some disadvantaged people (though it may also benefit others).

The next chapter will provide us with several ways to make more sense of the technological changes and ethical issues we have identified. As with many categorisation schemes, there may not be a single perfect way of doing so – think of ordering your bookcase or music collection, for instance – but what matters most is that we have a good way of working through them.

## Notes

1. Weiser (1991).
2. Crawford & Joler (2018).
3. Ubiquitous Computing (2018, June 2).
4. Not to be confused with the social media giant also known as Facebook.
5. Kitchin & McArdle (2016) and Laney (2001, February 6).
6. To borrow from the song title by Steely Dan.
7. Inspired by sources including Müller (2023), Computer History Museum (n.d.-b), Hillyer (2020, November 18), and Blackman (n.d.).
8. Asimov (1950).
9. Čapek & Čejková (2024).
10. Turing (1950).
11. Computer History Museum (n.d.-a).
12. See also the movie 'Hidden Figures' (2017), and Becker (2017, January 20).
13. Thompson (2019, June).
14. McCarthy, Minsky, Rochester, & Shannon (1955).
15. Wiener (1950).
16. Bassett (2019).
17. Weizenbaum (1976).
18. Iaccarino (2001).
19. Lighthill (1972).
20. NRC (2022).
21. Kuang (2021) and Walker (2005).
22. For an article on this development, see also, e.g., Froehlich (2004, December).
23. Mason (1986).
24. Kostrewski & Oppenheim (1979) and Froehlich (2004, December).
25. Baase (2013).
26. Leveson (1995).
27. For ongoing research on the digital divide, see Chapter 7 for an illustrative example.
28. Brown (2011).
29. Wasserman (2023, August 9).
30. Privacy International (n.d.).
31. Dapcevich (2024, January 12).
32. W3C (1999).
33. The case is officially known as *Maguire v Sydney Organising Committee for the Olympic Games*, No. 1112 (FCA August 3, 2000).
34. D'Onfro (2016, September 9) and Murphy (2013, March 12).
35. Natanson (2023).
36. PricewaterhouseCoopers (n.d.).
37. Fisher (2018, September 18).
38. For press coverage, see, e.g., Beaumont (2011, February 25).
39. Specifically, on Twitter, see: Cheong (2013) and Douglas (2011, August 9).
40. Ortiz-Ospina (2019).
41. Bozdag (2013).
42. For a primer, see What is deep learning? (2024, April 30).
43. Gabbatt (2011, February 17).

44. Borowiec (2016, March 9).
45. Sample (2020, January 13).
46. Haysom (2018, January 31).
47. Press coverage on Amazon's automation includes Dastin (2018, October 11).
48. Njoto (2024)'s PhD thesis studied the various the effects of AI and automation on gender bias and discrimination.
49. Examples of press coverage on Cambridge Analytica include: Gilbert & Ma (2019, April 24); Cadwalladr & Graham-Harrison (2018, March 17).
50. Officially: Regulation (EU) 2016/679 of the European Parliament and of the Council of 27 April 2016 on the protection of natural persons with regard to the processing of personal data and on the free movement of such data, and repealing Directive 95/46/EC (General Data Protection Regulation). (2016). OJ L, 119, 1–88. An accessible introduction can be found in the work of Wolford (2018).
51. European Commission (n.d.).
52. The foundational documents on these monumental moments in digital ethics include: Schloss Dagstuhl-Leibniz-Zentrum für Informatik GmbH. (n.d.) and Diakopoulos et al. (n.d.).
53. The FAccT conference website, in its current form, is at https://facctconference.org/
54. FAT ML. (n.d.).
55. IEEEXplore. (2014).
56. Artificial Intelligence, Ethics, and Society. (n.d.).
57. Asilomar AI Principles. (2017, August 11).
58. The IEEE Global Initiative 2.0 on Ethics of Autonomous and Intelligent Systems. (n.d.).
59. AI Principles Overview. (n.d.).
60. Marquez, N. (n.d.).
61. The authors and their colleagues work in one or more of these areas in Australia throughout the pandemic.
62. Interestingly, some of these examples are generated by ChatGPT (gpt-4o) itself, in response to a prompt on 'dot points on harms of Generative AI', generated 12 March 2025.
63. Press coverage on these harms include Singleton, Gerken, & McMahon (2023, October 6).
64. Pause Giant AI Experiments: An Open Letter (2023, March 22).
65. EU Artificial Intelligence Act (n.d.).
66. Artificial Intelligence Action Summit (2015).
67. The landmark AI Safety Report is a large-scale collaborative effort: see Bengio et al. (2025).

## References

AI Principles Overview. (n.d.). OECD.AI. Retrieved May 12, 2025, from https://oecd.ai/en/principles

Artificial Intelligence Action Summit. (2015). Elysee.Fr. Retrieved May 12, 2025, from https://www.elysee.fr/en/sommet-pour-l-action-sur-l-ia

Artificial Intelligence, Ethics, and Society. (n.d.). AAAI. Retrieved May 12, 2025, from https://aaai.org/conference/aies/

Asilomar AI Principles. (2017, August 11). Future of Life Institute. https://futureoflife.org/open-letter/ai-principles/

Asimov, I. (1950). I, Robot (The Isaac Asimov Collection Ed.). Doubleday.

Baase, S. (2013). 8.2 Case Study: The Therac-25. In A Gift of Fire: Social, Legal, and Ethical Issues for Computing Technology (4. ed). Pearson.

Bassett, C. (2019). The computational therapeutic: Exploring Weizenbaum's ELIZA as a history of the present. AI & SOCIETY, 34(4), 803–812. https://doi.org/10.1007/s00146-018-0825-9

Beaumont, P. (2011, February 25). The Truth About Twitter, Facebook and the Uprisings in the Arab World. The Guardian. https://www.theguardian.com/world/2011/feb/25/twitter-facebook-uprisings-arab-libya

Becker, K. (2017, January 20). When Computers were Human: The Black Women Behind NASA's Success. New Scientist. https://www.newscientist.com/article/2118526-when-computers-were-human-the-black-women-behind-nasas-success/

Bengio, Y., Mindermann, S., Privitera, D., Besiroglu, T., Bommasani, R., Casper, S., Choi, Y., Fox, P., Garfinkel, B., Goldfarb, D., Heidari, H., Ho, A., Kapoor, S., Khalatbari, L., Longpre, S., Manning, S., Mavroudis, V., Mazeika, M., Michael, J., & Zeng, Y. (2025). International AI Safety Report (No. arXiv:2501.17805). arXiv. https://doi.org/10.48550/arXiv.2501.17805

Bida, O. (2025). Orestbida/Cookieconsent [JavaScript]. https://github.com/orestbida/cookieconsent (Original work published 2020)

Blackman, R. (n.d.). How to Avoid the Ethical Nightmares of Emerging Technology. Harvard Business Review. Retrieved April 28, 2025, from https://hbr.org/2023/05/how-to-avoid-the-ethical-nightmares-of-emerging-technology

Borowiec, S. (2016, March 9). Google's AlphaGo AI Defeats Human in First Game of Go Contest. The Guardian. https://www.theguardian.com/technology/2016/mar/09/google-deepmind-alphago-ai-defeats-human-lee-sedol-first-game-go-contest

Bozdag, E. (2013). Bias in algorithmic filtering and personalization. Ethics and Information Technology, 15(3), 209–227. https://doi.org/10.1007/s10676-013-9321-6

Brown, M. (2011). Germany Lifts 17-year Ban on Demon-blaster Doom. Wired. https://www.wired.com/2011/09/germany-ban-doom/

Cadwalladr, C., & Graham-Harrison, E. (2018, March 17). Revealed: 50 Million Facebook Profiles Harvested for Cambridge Analytica in Major Data Breach. The Guardian. https://www.theguardian.com/news/2018/mar/17/cambridge-analytica-facebook-influence-us-election

Čapek, K., & Čejková, J. (2024). R.U.R. and the Vision of Artificial Life. The MIT Press.

Cheong, M. (2013). Inferring Social Behavior and Interaction on Twitter by Combining Metadata About Users & Messages [Thesis, Clayton School of Information Technology, Monash University]. https://doi.org/10.4225/03/58b5009d3726a

Computer History Museum. (n.d.-a). ENIAC. Computer History Museum. Retrieved April 28, 2025, from https://www.computerhistory.org/revolution/birth-of-the-computer/4/78

Computer History Museum. (n.d.-b). Timeline of Computer History. Retrieved April 28, 2025, from https://www.computerhistory.org/timeline/

Crawford, K., & Joler, V. (2018). Anatomy of an AI System: The Amazon Echo As An Anatomical Map of Human Labor, Data and Planetary Resources. AI Now Institute and Share Lab. http://www.anatomyof.ai

D'Onfro, J. (2016, September 9). Facebook's News Feed is 10 Years Old. This is How the Site has Changed. World Economic Forum. https://www.weforum.

org/stories/2016/09/facebooks-news-feed-is-10-years-old-this-is-how-the-site-has-changed/
Dapcevich, M. (2024, January 12). '90s Throwback: When Furbys Caused National Security Fears. Snopes. https://www.snopes.com//fact-check/nasa-furby-ban/
Dastin, J. (2018, October 11). Insight – Amazon Scraps Secret AI Recruiting Tool That Showed Bias Against Women | Reuters. Reuters. https://www.reuters.com/article/world/insight-amazon-scraps-secret-ai-recruiting-tool-that-showed-bias-against-women-idUSKCN1MK0AG/
Diakopoulos, N., Friedler, S., Arenas, M., Barocas, S., Hay, M., Howe, B., Jagadish, H. V., Unsworth, K., Sahuguet, A., Venkatasubramanian, S., Wilson, C., Yu, C., & Zevenbergen, B. (n.d.). Principles for Accountable Algorithms and a Social Impact Statement for Algorithms. FAT ML. Retrieved May 12, 2025, from https://www.fatml.org/resources/principles-for-accountable-algorithms
Douglas, T. (2011, August 9). Social Media's Role in the Riots. BBC News. https://www.bbc.com/news/entertainment-arts-14457809
EU Artificial Intelligence Act. (n.d.). EU Artificial Intelligence Act. Retrieved May 12, 2025, from https://artificialintelligenceact.eu/
European Commission. (n.d.). Legal Framework of EU Data Protection. European Commission. Retrieved May 12, 2025, from https://commission.europa.eu/law/law-topic/data-protection/legal-framework-eu-data-protection_en
FAT ML. (n.d.). Home. FAT ML. Retrieved May 12, 2025, from https://www.fatml.org/
Fisher, A. (2018, September 18). Google's Self-Driving Cars: A Quest for Acceptance. Popular Science. https://www.popsci.com/cars/article/2013-09/google-self-driving-car/
Froehlich, T. (2004, December). A Brief History of Information Ethics [Text.Article]. BiD: Textos Universitaris de Biblioteconomia i Documentació. https://bid.ub.edu/13froel2.htm
Gabbatt, A. (2011, February 17). IBM Computer Watson Wins Jeopardy clash. The Guardian. https://www.theguardian.com/technology/2011/feb/17/ibm-computer-watson-wins-jeopardy
Gilbert, B., & Ma, A. (2019, April 24). Facebook Understood How Dangerous the Trump-Linked Data Firm Cambridge Analytica Could be Much Earlier Than It Previously Said. Here's Everything That's Happened Up Until Now. Business Insider. https://www.businessinsider.com/cambridge-analytica-a-guide-to-the-trump-linked-data-firm-that-harvested-50-million-facebook-profiles-2018-3
Haysom, S. (2018, January 31). People are Using Face-swapping Tech to Add Nicolas Cage to Random Movies and What is 2018. Mashable. https://mashable.com/article/nicolas-cage-face-swapping-deepfakes
Hillyer, M. (2020, November 18). Here's How Technology has Changed the World Since 2000. World Economic Forum. https://www.weforum.org/stories/2020/11/heres-how-technology-has-changed-and-changed-us-over-the-past-20-years/
Iaccarino, M. (2001). Science and ethics. EMBO Reports, 2(9), 747–750. https://doi.org/10.1093/embo-reports/kve191
IEEEXplore. (2014). 2014 IEEE International Symposium on Ethics in Science, Technology and Engineering. IEEEXplore. https://ieeexplore.ieee.org/xpl/conhome/6883275/proceeding
Kitchin, R., & McArdle, G. (2016). What makes big data, big data? Exploring the ontological characteristics of 26 datasets. Big Data & Society, 3(1), 2053951716631130. https://doi.org/10.1177/2053951716631130

Kostrewski, B. J., & Oppenheim, C. (1979). Ethics in information science. Journal of Information Science, 1(5), 277–283. https://doi.org/10.1177/016555157900100505

Kuang, C. (2021). How Design Saved UX. Google Design. https://design.google/library/user-friendly

Laney, D. (2001, February 6). 3D Data Management: Controlling Data Volume, Velocity, and Variety. Application Delivery Strategies – Meta Group. https://web.archive.org/web/20120304154148/blogs.gartner.com/doug-laney/files/2012/01/ad949-3D-Data-Management-Controlling-Data-Volume-Velocity-and-Variety.pdf

Leveson, N. (1995). Medical devices: The therac-25*. In SafeWare: System Safety and Computers. Addison-Wesley. http://sunnyday.mit.edu/papers/therac.pdf

Lighthill, J. (1972). Artificial Intelligence: A General Survey. Science Research Council. https://www.chilton-computing.org.uk/inf/literature/reports/lighthill_report/p001.htm

Maguire v Sydney Organising Committee for the Olympic Games, No. 1112 (FCA August 3, 2000). https://www.austlii.edu.au/cgi-bin/viewdoc/au/cases/cth/FCA/2000/1112.html

Marquez, N. (n.d.). Misinformation – Get the Facts: What is Misinformation? UC Irving Libraries. Retrieved May 12, 2025, from https://guides.lib.uci.edu/Misinfo/WhatIsMisinfo

Mason, R. O. (1986). Four ethical issues of the information age. MIS Quarterly, 10(1), 5–12. https://doi.org/10.2307/248873

McCarthy, J., Minsky, M. L., Rochester, N., & Shannon, C. E. (1955). A Proposal for the Dartmouth Summer Research Project on Artificial Intelligence. http://jmc.stanford.edu/articles/dartmouth/dartmouth.pdf

Müller, V. C. (2023). The history of digital ethics. In C. Véliz (Ed.), Oxford Handbook of Digital Ethics. Oxford University Press. https://doi.org/10.1093/oxfordhb/9780198857815.013.1

Murphy, S. (2013, March 12). The Evolution of Facebook News Feed. Mashable. https://mashable.com/archive/facebook-news-feed-evolution

Natanson, E. (2023). Hyper-Personalization Is Already Here – Its Future Is Even More Cutting-Edge. Forbes. Retrieved May 12, 2025, from https://www.forbes.com/sites/eladnatanson/2023/06/01/hyper-personalization-is-already-here---its-future-is-even-more-cutting-edge/

Njoto, S. (2024). Decoding Discrimination: Unraveling Gender Bias in Semi-Automated Recruitment [University of Melbourne]. https://trove.nla.gov.au/work/256304222?keyword=Decoding%20Discrimination%3A%20Unraveling%20Gender%20Bias%20in%20Semi-Automated%20Recruitment

NRC. (2022). Backgrounder on the Three Mile Island Accident. NRC. https://www.nrc.gov/reading-rm/doc-collections/fact-sheets/3mile-isle.html

Ortiz-Ospina, E. (2019). The Rise of Social Media. Our World in Data. https://ourworldindata.org/rise-of-social-media

Pause Giant AI Experiments: An Open Letter. (2023, March 22). Future of Life Institute. https://futureoflife.org/open-letter/pause-giant-ai-experiments/

PricewaterhouseCoopers. (n.d.). Making Sense of Bitcoin, Cryptocurrency and Blockchain. PwC. Retrieved May 12, 2025, from https://www.pwc.com/us/en/industries/financial-services/fintech/bitcoin-blockchain-cryptocurrency.html

Privacy International. (n.d.). Micro-targeting. Privacy International. Retrieved May 5, 2025, from https://privacyinternational.org/learn/micro-targeting

Regulation (EU) 2016/679 of the European Parliament and of the Council of 27 April 2016 on the protection of natural persons with regard to the processing of personal data and on the free movement of such data, and repealing Directive 95/46/EC (General Data Protection Regulation). (2016). OJ L, 119, 1–88.

Sample, I. (2020, January 13). What are Deepfakes – And How Can You Spot Them? The Guardian. https://www.theguardian.com/technology/2020/jan/13/what-are-deepfakes-and-how-can-you-spot-them

Schloss Dagstuhl-Leibniz-Zentrum für Informatik GmbH. (n.d.). Dagstuhl Seminar 16291: Data, Responsibly. Retrieved May 12, 2025, from https://www.dagstuhl.de/en/seminars/seminar-calendar/seminar-details/16291

Singleton, T., Gerken, T., & McMahon, L. (2023, October 6). How a Chatbot Encouraged a Man Who Wanted to Kill the Queen. https://www.bbc.com/news/technology-67012224

The IEEE Global Initiative 2.0 on Ethics of Autonomous and Intelligent Systems. (n.d.). IEEE Standards Association. Retrieved May 12, 2025, from https://standards.ieee.org/industry-connections/activities/ieee-global-initiative/

Thompson, C. (2019, June). The Gendered History of Human Computers. Smithsonian Magazine. https://www.smithsonianmag.com/science-nature/history-human-computers-180972202/

Turing, A. M. (10 1950). I. – Computing machinery and intelligence. Mind; a Quarterly Review of Psychology and Philosophy, LIX(236), 433–460. https://doi.org/10.1093/mind/LIX.236.433

Ubiquitous Computing. (2018, June 2). https://web.archive.org/web/20180602090007/http://www.ubiq.com/hypertext/weiser/UbiHome.html

W3C. (1999). Web Content Accessibility Guidelines 1.0. https://www.w3.org/TR/WAI-WEBCONTENT/

Walker, J. S. (2005). Three Mile Island: A Nuclear Crisis in Historical Perspective (1. Paperb. Pr). University of California Press.

Wasserman, T. (2023, August 9). This Is the World's First Banner Ad. Mashable. https://mashable.com/archive/first-banner-ad

Weiser, M. (1991). The computer for the 21st century on JSTOR. Scientific American, 265(3), 94–105.

Weizenbaum, J. (1976). Computer Power and Human Reason; From Judgement to Calculation. W. H. Freeman and Company.

What is deep learning? (2024, April 30). McKinsey & Co. https://www.mckinsey.com/featured-insights/mckinsey-explainers/what-is-deep-learning

Wiener, N. (1950). The Human Use of Human Beings: Cybernetics and Society. Houghton Mifflin.

Wolford, B. (2018, November 7). What is GDPR, the EU's New Data Protection Law? GDPR.Eu. https://gdpr.eu/what-is-gdpr/

# 3

# *What Is the Problem with Tech?*

As we have hinted, categorising the myriad issues in digital ethics is no easy feat. Again, think of your music collection. You could classify all *'rock'* music under one section; your friend who is a music enthusiast might further divide it into *'glam rock'*, *'hard rock'*, *'indie rock'*, and so forth. Some artists span multiple genres: Think of Taylor Swift and Beyoncé, who are both in *'pop'* and *'country'*. Where would we start? The same applies to the many and various ethical problems and issues we have so far touched on.

## Our Taxonomy

In current academic literature – computer science, philosophy, law, the social sciences, and more – there is general agreement on the broad topic areas of digital ethics. For example, it is widely agreed that digital misinformation, psychological profiling and microtargeting by platforms, automated decision-making, and the uncontrolled generation of realistic images and text are pressing moral concerns. However, there is no single authoritative taxonomy of the issues in digital ethics, and there are, furthermore, various overlaps between these issues.[1] For the sake of argument, an artificial intelligence (AI) algorithm could be described as *unfair* either because its statistical and mathematical underpinnings are flawed or because the outcomes it produces when used in certain ways are unfair (without critiquing the math).

Thus, in this chapter, we have considered several ways to classify the various types of issues prevalent in digital ethics. We give the following categorisation, inspired by several sources, including:

- our experiences as educators;
- pre-existing categorisations in academic literature across disciplines;
- categorisations used by policymakers such as the OECD; and even
- media stories and common knowledge of ethical issues as encoded in modern Generative AI systems.[2]

The following sections present, in no particular order, a non-exhaustive categorisation of major concerns in digital ethics, as well as relevant examples

DOI: 10.1201/9781003393511-3

(including those from Chapter 2), and potential countermeasures to be aware of. We must stress that it is impossible to cover *all* digital ethics issues in this book; inevitably, we will leave some out.

## Privacy and Data Governance

### Privacy

Firstly, let's take an issue affecting many areas of everyday life, beyond the digital – that of **privacy**. Privacy, a complex idea, intuitively means:

> a state in which one is not observed or disturbed by other people[3]

Privacy is a concern that grew alongside the developments in computing in the past few decades: In fact, Mason's *PAPA* principles (*Privacy, Accuracy, Property, and Accessibility*, as per Chapter 2) listed Privacy first and foremost.[4] (Note that the definition of privacy will be re-examined in Chapter 5, when it is discussed as an ethical principle.)

In the context of digital technology, we can roughly take privacy protection to relate to the protection of personal data from unauthorised or unwanted access. We might also take privacy to include not being disturbed by others online, such as keeping our contact details safe to avoid spam and unwanted contact. (As we have discussed earlier though, this latter part overlaps with what we term 'safety', which will be expanded later on in this chapter.)

Some privacy threats are due to the technology we use, for example, accidentally giving too much away on social media platforms. Other threats are from technologies which may use our data for secondary purposes, such as extra information collected by apps which may cause us to be the target of more relevant ads. And other threats sadly are out of control, such as our data held by organisations which may be breached by malign actors.[5]

*Examples in Chapter 2*: The *Furby* toy and *social media sites*, which can collect and store our data.

*Considerations to be aware of*[6]:

- Being parsimonious with our data by considering how much we share and give away (e.g., boycotting apps which require an inordinate amount of data in the registration process).
- Being aware of one's potential involvement in data breaches through websites such as Have I Been Pwned[7] or Mozilla Monitor.[8]
- 'Cleaning' our social media presences.
- Adopting privacy-preserving technologies such as ad and web tracking blockers, utilising web browser and smartphone features to promote privacy.

## Data Governance

In the context of data privacy (ensuring our data is kept away from prying eyes), the way data is kept and governed plays an important role (including how the data vouchsafed in the first place!)

Data governance has been discussed in multiple contexts, including:

- Law: e.g., with the European Union's General Data Protection Regulation (GDPR).[9]
- Philosophy: and the link between governance and digital ethics.[10]
- Computing: e.g., the latter three of Mason's PAPA principles.[11]
- Industry: e.g., McKinsey's data-governance model and principles.[12]

A working ensemble definition of **data governance**, assembled from different sources, is:

> … the principles/rules governing data – including its access, accountabilities, decision-making, and enforcement – to ensure data is accurate and risk-free.[13]

At first glance, we might think that this is solely the domain of the people or organisations that manage or keep our data. For one, it links directly to the issue of data privacy seen earlier: especially if a privacy breach is due to sloppy management of data, say. One can argue that traditionally, these are legal risk/compliance issues for organisations.

We may have been made more aware of data governance after the introduction of the EU GDPR, which brings to the forefront obligations and practices such as:

> data protection is something [the data processor] … now has to consider whenever [it has] … anything with other people's personal data …
>
> people have the right to see what personal data [the data processor has] … about them and how [they are] … using it …
>
> [the data processor is] … also required to quickly communicate data breaches to … data subjects[14]

However, data governance is a (digital) ethical issue because it concerns the impact on us as **data subjects** – in other words, we are people who could be identified by our personal data.[15] Some questions to pique your interest are: *Who owns your data on the 'cloud'?*, *What happens to your cookies and the consent you gave (or otherwise) to accept them?*, *How do large language models and Generative AI use your data in their models?*, and also *What happens to your social media profile after you pass away?* These, and more, are important things to consider as we are increasingly dependent on, and as we increasingly adopt, new technologies which create new data on an unprecedented level.

*Examples in Chapter 2: social media sites, Cambridge Analytica* scandal, *ChatGPT* use.

*Considerations to be aware of*:

- Understanding how data is used and gaining an awareness of the issues involving data use/reuse.
- Adopting data parsimony, or **Datensparsamkeit**[16] in German: the principle of collecting (and providing) the least amount of data required to perform a task.
- Following legal requirements (such as the GDPR and Australian Privacy Principles[17]) in the collection and use of others' data.

## Fairness, Accountability, and Transparency

The three topics of Fairness, Accountability, and Transparency (F.A.T.) have been the subject of academic interest since the 2010s. As mentioned in Chapter 2, the *Fairness, Accountability, and Transparency in Machine Learning* (FAT/ML, now FAccT) conference was one of the pioneers in this topic of inquiry.

Starting out with a focus on CS/ML and law,[18] the F.A.T. topics are now being researched by social scientists, IS and HCI practitioners, and researchers in the humanities.

We will now unpack these three interrelated areas below.

### Fairness

An aim for algorithmic **fairness**, or fairness in fields such as ML,[19] is to understand when "… disparities are harmful, unjustified, or otherwise unacceptable, and to develop interventions to mitigate such disparities".[20]

Algorithmic fairness is intertwined with another ethical concern in the following section, that of equity of outcomes. In this section, we focus on the issues in the algorithms themselves.

Issues of fairness range from the potential lopsidedness or bias found in existing datasets used to train algorithms, to the very definitions of fairness used to judge or evaluate algorithms. The latter point is interesting; researchers note that there is a plurality of different definitions of fairness, which can contradict one another.[21] About 21 such definitions have been identified by the researcher Arvind Narayanan![22]

*Examples in Chapter 2*: The *Fairness, Accountability, and Transparency in Machine Learning* initiative in bringing these issues to the forefront, such as *gender bias* in algorithms.

*Considerations to be aware of*:

- Competing and sometimes incompatible definitions of 'fairness' in algorithms can lead to different 'metrics' of how fair a system is.
- The role of biased data in training algorithms to make decisions.
- Potential cherry-picking of fairness measures in justifying decisions.

### Accountability

We can define **accountability** here as holding parties responsible for ethical and legal outcomes of AI, among other algorithmic and technological systems.

For one, what happens if the outcome of a decision-making system is detrimental to a **decision subject** (simply put, the person or people affected by a decision)? There should be opportunities for those impacted to seek *redress for, recourse to, or contest,* these outcomes.[23] Article 22 in the EU GDPR even provides a right *against* automated decision-making for an individual.[24]

When decisions made by these systems affect the public sphere, such as in law enforcement and administrative decision-making,[25] what measures are there to ameliorate the risk? The example of the Australian Robodebt scheme (in the Preface) is a cautionary tale of issues with accountability in such a *high stakes* scenario.[26]

It is important to also note that for complex systems with 'many moving parts', responsibility may be attributed to the wrong person when things go wrong.[27] On the other hand, the human users of technological systems may underestimate the fallibility of these systems, such as in the landmark case of a lawyer whose use of ChatGPT led to non-existent 'hallucinated' cases being cited (causing public embarrassment for the lawyer concerned).[28]

*Examples in Chapter 2*: The *Three Mile Island* nuclear accident; the *Therac-25* failure, accountability of *COVID-19 check-in* systems.

*Considerations to be aware of*:

- Adequate legal safeguards need to be in place to protect *decision subjects* (e.g., the ability to challenge the decisions).
- A proportional degree of trust[29] in technological systems must be based on evidence and evaluation of their capabilities.
- Adequate auditing, monitoring, reporting, and compliance mechanisms to make sure systems are doing the right thing.

### Transparency

Finally, under the banner of F.A.T., the issue of transparency is a multidimensional issue in the realm of digital ethics.

**Transparency**, at a big picture perspective, refers to the overall openness of a technological system's development, deployment, and usage practices. This includes everything from the planning stages to the implementation; and encompasses both the code and the data.

Consider a new device and its operating system. A best practice mechanism to ensure transparency is to make its code and designs **open source**[30]: In other words, its inner workings are freely available for others to examine and audit. Often, due to issues such as intellectual property protection, this is not possible: an alternative is to invite external experts to audit them.[31]

Transparency – at a more 'micro' level – can also refer to making the inner workings of an algorithmic system understandable. On the one hand, one could use simple algorithms to make decisions or predictions **interpretable** – the degree of 'how well a human could understand the decisions in the given context'.[32] One way is to use simple decision trees – collections of rules such as 'If X, then do Y' – rather than complex neural network algorithms.[33]

On the other hand, given the technical complexity of a particular system, another approach is to make the *rationale* behind the predictions or decisions clear. **Explainability** concerns the ability of an AI (or other technological) system to "reveal … underlying causes to its or another agent's decision making".[34] Explainability is a useful feature that promotes fairness (one knows why a decision is made) and accountability (one knows how the decision can be challenged).

*Examples in Chapter 2*: Open Source Software such as *Linux*.

*Considerations to be aware of*:

- The balancing of intellectual property, trade secrets, or security, with the need for transparency.
- How useful the explanations are for allowing the decision subject to make a meaningful change. This involves more than mere 'lip service' or 'nominal' transparency – such as providing vague reasons or generic explanations of the algorithm – that doesn't benefit the subject.[35]
- Ensuring that the entire system – and not just cherry-picked individual components – is considered holistically in transparency-promoting efforts.

## Equity and Accessibility

In the past subsection, we've mainly been discussing the vagaries and pitfalls of a system by taking an algorithmic perspective. However, issues such as biased selection algorithms, or non-accountable 'black boxes', have real-world impact. This is where the twin themes of equitability and accessibility come in.

## Equitability

**Equitability** (or **equity**) is about making sure that the use and outputs of technological systems provide the 'same access to opportunities'[36] to its different users.

Now, although *equity* is related to *equality*, they are not necessarily the same thing. A frequently quoted metaphor is one of multiple people trying to look past a picket fence, each with different characteristics (Figure 3.1). Equality, in this analogy, could be seen as giving everyone the *same* footstool to boost their height irrespective of their needs, which doesn't consider their individual circumstances. Equity, by contrast, entails making sure everyone's needs are catered for: Some might need footstools, some others ladders, and some others might need assistance to climb the fence.

Issues of *in*equity are prevalent in the literature and popular press. (More discussion can be found in Chapter 7). Here are some examples:

- AI tools that recommended less medical care for African-Americans compared to White-Americans.[37]
- Hiring algorithms that exhibit bias against women and parents.[38]

**FIGURE 3.1**
A metaphor illustrating the difference between equity and equality. (*Image (IISC & Maguire, 2016) is by Interaction Institute for Social Change, Artist: Angus Maguire, provided under the Creative Commons CC BY-SA 4.0 - Attribution-ShareAlike 4.0 International - license*)

- Predictive policing and criminal recidivism prediction systems that entrench bias against minority groups.[39]
- Generative AI models that exhibit biased or stereotyping tendencies across people from different racial and religious groups.[40]

The **digital divide** is another area in this field that pre-dates the modern rise of (potentially biased) algorithmic decision-making technologies. According to the OECD, the digital divide is:

> ... the gap between individuals, households, businesses and geographic areas at different socio-economic levels with regard to their opportunities to access information and communication technologies (ICTs).[41]

Scholars have since identified multiple levels of the digital divide: ranging from the divide in access to technology and Internet connectivity, to the divide in skills and usage barriers, to the divide in the equity of outcomes.[42]

Thus, from our coverage above, one can see how the digital divide (in terms of access and outcome inequity) coupled with the adverse effects of decision-making algorithms, could harm people in minoritised groups. Some researchers posit that due to the scale and speed of AI and automated decision making, they could perpetuate further disadvantage on those already digitally excluded.[43]

*Examples in Chapter 2: Gender bias* in hiring, history of *discrimination in the technology sector*, potential for *Generative AI* to embed biases, *COVID-related technologies and policies* which can exacerbate the digital divide (e.g., from contact tracing apps to cashless economy technologies).

*Considerations to be aware of*:

- The potential for *biases* impacting populations who are using digital technology or are otherwise the *data subject* (e.g., affected by a decision-making process).
- Not introducing technology purely for the sake of innovation (so-called *technosolutionism*), especially if the technology has the possibility of doing more *harm* than good (see the proposed EU AI Act that prohibits certain very high-risk AI technologies).[44]
- Performing thorough audits of technology before and after its adoption, with continuous monitoring to anticipate potential issues, including bias and widening of the *digital divide* (including affordability and concerns with digital literacy).
- Considering the widespread impact on *existing users* before introducing a new requirement for technology.[45]
- Being aware of the technology trend towards *planned obsolescence* (more in *Environment and Sustainability* later).

## Accessibility

As mentioned, issues of equity are commonly linked to issues of accessibility. **Accessibility** is defined in the HCI literature as:

> ... the degree to which an interactive product is accessible by as many people as possible. A focus is on people with disabilities.[46]

Yet accessibility also goes beyond simply *catering for those with disabilities*. The analogy of the elevator helps here: While it is true that it helps wheelchair users, it also helps people who are pressed for time, carrying heavy packages, or wanting to safely transport toddlers.

The same can be said for digital technologies such as subtitles and dictation (or speech-to-text systems), and voiceovers and narration (text-to-speech systems). It is true these technologies benefit people with accessibility requirements, but people without disabilities could additionally benefit from their introduction. To take just a couple of examples, using subtitles can help us watch a film satisfactorily in a loud environment, and voiceovers could help us proofread our emails more effectively.

*Examples in Chapter 2: Web Content Accessibility Guidelines*, the *Maguire vs Sydney Organising Committee for the Olympic Games (SOCOG)* case.

*Considerations to be aware of*:

- Ensuring that the introduction of technology does not cause *accessibility* issues that affect both people with and without disabilities.
- Always providing support and other alternative mechanisms (e.g., human assistance, rather than purely depending on AI) to promote equitable use.
- Adhering to laws and regulations, as well as best-practice principles (such as the Web Content Accessibility Guidelines, or WCAG) regarding accessibility.
- Fostering co-design *with* people with lived experience of disability when creating, designing, and evaluating technological products (see also Chapter 7).

## Security and Safety

Safety is a repeated theme in the history of computing and technology (Chapter 2) as well as popular fiction (ranging from HAL in *2001: A Space Odyssey* to the *Terminator* franchise). We need to be able to have confidence in AI systems to function reliably, safely, and ethically.

We can briefly divide this topic into two, again interrelated, aspects.

### Psychological Harm and Epistemic Concerns

Let us start off with one aspect that was alluded to in our case study in the Introduction. To recap:

> Facebook allowed teenagers to be exposed to content about unrealistic body image, which affected their mental health and caused them to be vulnerable.[47]

We could classify this sort of harm as **psychological harm**.[48] In the example above, the *recommender systems* in charge of providing a curated feed for the social media user inadvertently promoted content about unrealistic body image.

Facebook's algorithms, of course, are not the only problematic technology. Consider flawed chatbots which provide bad health advice harming vulnerable users, or even worse, cause them to harm themselves.[49]

Related to this is the effect of technologies that harm the capacity of a user as a 'knower' of justified knowledge. We term these 'epistemic concerns'. An example is filter bubbles[50] (Chapter 2) and more insidiously, the *echo chamber*[51] effect on social media platforms, which cause opposing viewpoints to be actively discredited. Examples in recent memory include the active propagation of vaccine disinformation on social media during the COVID-19 pandemic[52]; the manipulation of political viewpoints via 'microtargeting' in the case of Cambridge Analytica[53] (Chapter 2); and the potential of website architecture to 'nudge' the reader to believe certain narratives based on the news they read.[54]

*Examples in Chapter 2*: *Social media sites* and filter bubbles, *Cambridge Analytica* scandal, *COVID-19 misinformation*, *ChatGPT's harms* to vulnerable users.

*Considerations to be aware of*:

- Understanding how social media sustains our attention, and the *potential harms of recommendation and filtering algorithms.*
- Identifying trusted sources of news to follow, rather than blindly accepting 'news' from social media.
- Using social media and related technologies *mindfully*, e.g., in moderation or with 'social media diets'.
- Understanding the *limitations of chatbots and other technologies in healthcare*, such as in mental health therapy[55]

### Physical Safety

Besides direct harms caused by the examples above (such as dangerous health practices because of a belief in health disinformation online), Chapter 2 has briefly demonstrated a myriad of issues directly affecting the physical safety of the users of technology as well as innocent bystanders.

A quintessential example is the deployment of *autonomous vehicles – self-driving cars* – and their safety concerns. The ones reported in the news range from the inconvenient (a person being trapped in a self-driving car for a few minutes[56]) to more tragic instances (including potential injury and loss of life[57]). Beyond self-driving cars, other fatal incidents in the past decades range from an accident with industrial robots, documented as early as the late 1970s,[58] to the grave safety issues afflicting the Boeing 737 MAX airplane which resulted in catastrophic plane crashes.[59]

*Examples in Chapter 2: Three Mile Island, Therac-25.*

*Considerations to be aware of:*

- Evaluating the safety of new technologies before adopting them; again, *avoiding technosolutionism and hype,* especially if potential harms can outweigh the benefits.
- Following best practices, ensuring compliance, and managing risk before a new piece of technology is deployed.

## Environmental Issues

Most of our discussion thus far dealt with the impact of badly-thought-out or badly implemented technologies on us, humans, or our societies.

Yet the scholarship on digital ethics has also more recently recognised the harms of technology to our environment, ecosystems, and (non-human) animals. The following three subsections will outline these concerns.

### Energy Consumption and Carbon Emissions

A great deal of natural resources is used in our pursuit of bigger and better technology. First, and most obvious, of all concerns tech's energy usage. The data centres housing key systems and data worldwide have been estimated to consume between 1% and 2% of global electricity demand, based on International Energy Agency and other industry estimates.[60] The percentage of data centre power consumption relative to developing countries (Australia and the USA) is about 5%.[61]

These figures are set to grow thanks to the rise of Generative AI systems. Scholars have estimated that the training process for GPT-3 required the use of 1287 MWh, with emissions of 552 tonnes of $CO_2$.[62] This is equivalent to the energy needed to charge over 40 million smartphones![63]

This does not consider the carbon footprint and amount of energy subsequently used in the process of generating media with Generative AI. One study revealed that the generation of two images approximates the amount of energy needed to charge two smartphones, on average.[64] Technologists argue that using a large language model to find information 'likely cost 10 times more than a standard keyword search'.[65]

*Example in Chapter 2: ChatGPT* (as an example of the widespread use of Generative AI).

*Considerations to be aware of*:

- Being aware about how much energy is used by AI and other technologies.
- Understanding the trade-off between more advanced technologies versus existing technologies in sustainability terms.

### Resource Depletion, eWaste, and Planned Obsolescence

As with any human-made object, the development of hardware used to power the latest technology – from smartphones, to laptops, to server farms – requires the use of raw materials. Technological devices are notorious for their use of rare earth metals: from parts as small as light emitting diodes (LEDs), to assemblages that are our modern-day smartphones and electric vehicles.[66] The extraction of these rare earth elements contributes to the release of toxic chemicals that harm humans, animals, and ecosystems.[67]

To compound this problem, the culture of consumerism has led to the attraction of 'bigger and better' tech. Many of us are guilty of upgrading our computers and devices to cope with the increasing demands of speed and convenience, intentional or otherwise: Think of our phones no longer able to support the newest apps we might need to do our work, or becoming more sluggish/laggy every passing day. This is known as the push towards **planned obsolescence**, mentioned as early as the 1980s by Jeremy Bulow:

> the production of goods with uneconomically short useful lives so that customers will have to make repeat purchases.[68]

Consider this: If and when a device breaks down, do we tend to get a replacement, or attempt to repair it? One might argue that the problem with many modern devices is that they are hard to repair. Compare the difficulty in replacing the battery of one's smartphone with the difficulty in replacing the battery in an older phone from the 2000s, say the Nokia 3210. Also, consider that in some cases, an affordable repair is hard to come by: Repairers are constrained by the official repair know-how, proprietary tools, and costly spare parts.[69] This difficulty-to-repair is not only limited to smart devices: Even tractors are not spared![70]

The Australian Government's Productivity Commission indicated that

> there are significant and unnecessary barriers to repair for some products …

> … The rise in tech-enabled products means that much of the information required to diagnose a fault is digital, embedded into the product itself and held behind 'digital locks', requiring passwords or special tools to bypass[71]

Hearteningly, the **right to repair** movement is gaining traction,[72] with various laws being passed globally to support the consumer in sustainable repair of technology, rather than needing to fork out more money to buy new ones.[73] An example of a product that is easily maintained is the custom-built desktop PC: Parts can be swapped out as needed with compatible replacements, with minimal do-it-yourself training.

*Example in Chapter 2: Smartphone development* and use.

*Considerations to be aware of:*

- Understanding the motivations behind *the right to repair movement,* and the blockers to sustainable use of technology.
- *Avoiding technosolutionism* or the need for 'the latest and greatest' technologies, if an older one can suffice.

## Ecological Harms and Harms to Non-Human Life

While we have covered harms to human life and safety in the past chapter, it will be remiss of us if we did not emphasise harms to natural ecosystems, flora, and animals. These harms are sometimes or often overlooked in technology discussions.

Pollutants are noted as the result of activities done in pursuit of technology (such as rare earth mining as previously discussed). However, accidents can also happen, which inadvertently release pollutants on a large-scale and damage ecosystems. Fortunately, the Three Mile Island incident which we mentioned earlier has caused 'negligible effects on the physical health of individuals or the environment',[74] but larger incidents could result in foreseeably immense damage.

In recent times, the use of AI and other technologies in agriculture and similar areas, say, to automate farms in the name of improving their productivity, raises concerns about animal welfare.[75] Animals may also be indirectly harmed as a side effect of benevolent technology: Consider the hacking of technologies that are (ironically) used for protecting wildlife and other protected species.

*Example in Chapter 2: Three Mile Island* accident (hypothetically); general *cybersecurity* threats.

*Considerations to be aware of:*

- The accidental harms caused to the environment, plants, and animals, if technology were to go wrong.
- *Unforeseen consequences of technology* which might backfire, sometimes at the expense of non-human creatures.

## Society

To end this chapter, we highlight some of the issues of digital ethics regarding society at large. Some of these issues have come up in the discourse on, say, the proliferation of social media. Others are not new; they have been the subject of discussion since the introduction of technologically enabled automation, digitisation, and invention, as pointed out in Chapter 2.

### Effects to Democracy and The Rule of Law

As the global elections in the 2010s and 2020s have shown, modern technologies such as social media could influence people *en masse*, for better or for worse.

A cautionary landmark example is the effect of Cambridge Analytica in influencing elections, given the amount of information we have on a personal level being accessible on social media. Misinformation and disinformation are both noticeable during election campaigns[76]; and the rapid growth in Generative AI technology (from text generation to deepfakes) is accelerating these issues.

*Examples in Chapter 2: Cambridge Analytica* vis-à-vis the rise of *social media; Generative AI* and *deepfakes.*

*Considerations to be aware of*:

- Awareness of the impact of misinformation and disinformation in the public sphere.
- Knowledge of how to spot misinformation and disinformation.

### Economic Harms

Especially with the creation of advanced AI technologies and automation, job loss is a common theme echoed by the general public, as these new pieces of tech have the ability to displace workers and render current jobs at risk, while introducing their own set of challenges.

Interestingly, events such as the Luddite movement – oftentimes confused with being *technophobic* or blindly opposing changes in technology – highlight the precarity and potential displacement of workers when faced with new technologies, such as (digital) automation.[77]

Take, for instance, rideshare and other **gig economy**[78] apps, such as Uber/UberEats, Grab, Didi, and Doordash, to name a few. Rideshare apps and gig work are double-edged swords: Their debut disrupted the taxi industry in the 2010s with early claims of empowering drivers who adopt these apps,[79] while concerns such as safety and welfare of gig workers are still the subject of ongoing debate.[80]

The introduction of AI further introduces threats to the economic prospects of the general public. According to an Australian study, "1-in-10

Australian jobs are at risk of being automated or replaced by artificial intelligence".[81] The study initially found that, amongst their sample of Australian workers in 2022, 'very few worry about the currency of their skills' when faced with modern AI technologies, which sounded optimistic at the time of the study.

ChatGPT and its many competitors have, arguably, turned this picture around. Concerns about job security and deskilling[82] are noteworthy, with a 2023 Ipsos study indicating that approximately 36% of their respondents opine that in 5 years, "AI will replace … [their] current job", while about 57% think AI will "change how [they] … do [their] … current job".[83]

Besides the uncertainty in one's income, technology could unfortunately also play a role in making us lose our hard-earned money. Assisted by technology such as Internet-powered phone systems,[84] scammers can impersonate another individual with deepfaked audio and/or video, to convince their victims to part with large sums of money.[85] Various other cryptocurrency investment scams and malicious ransomware attacks (where precious data is held hostage until the victim pays exorbitant sums of money) are just another few examples.

*Examples in Chapter 2: Online delivery* (popular during the COVID-19 pandemic), *cryptocurrency, Generative AI* and *deepfakes* in scams, *worms and viruses* for cyber-theft.

*Considerations to be aware of*:

- Impact of technology on current jobs, including redundancies and excessive/undue surveillance or micromanagement.
- Realising the need to upskill the current workforce in order to master AI and new technologies, while *avoiding hype or panic* associated with them.
- Knowledge to identify and counter threats to financial safety (including malware and financial cybercrime).
- A *healthy dose of scepticism and critical thinking* to counter technologically-enabled scam/fraud methods which play on human emotions (such as fear and worry).

## Overdependence and Loss of Intrinsic Value

Lastly, we shall note a possible coming threat to society, in the form of increasing overdependence on digital tech and the stifling of humanistic qualities (e.g., creativity and innovation).

People everywhere today are engrossed in their smartphones, their attention absorbed by endless streams of notifications. Families at dining tables complain about not having 'genuine' communication when everyone is texting or watching. (Anecdotally, kids are increasingly able to bypass screen time restrictions on their smart devices!) Even educators have the same complaints, albeit in a lecture setting.[86]

This raises the concern of potential *overdependence* on tech: from overuse of our devices, to online gaming addiction, to obsessions with notifications on social media. If left unchecked, they can be causes for concern. Concerns such as *nomophobia*[87] (short for 'no mobile phone phobia') and video game addiction[88] are amongst issues[92] to be cognisant of.

Closely related to our prior discussion on deskilling, technology has the potential of instilling bad work habits, such as taking unjustified shortcuts, unquestioning complacency, and assuming technology is 'right'.[89] With the advent of Generative AI, we may face an unhealthy future dependence on such tools to do our work for us, encouraged by how easy it is to deploy and use these technologies with minimal entry barriers.

As seen the previous section, deskilling in one's job is one thing. Another threat is when humanistic pursuits, enmeshed in technology, lose their *intrinsic* good, merely to become *instrumental* acts.[90] Intrinsic acts are those that are valuable in themselves (such as producing a watercolour painting, or playing a good round of football), while instrumental acts are valuable insofar as they can help us achieve something else (such as drinking merely to quench one's thirst). An example is the lopsided use of an AI product to displace or compete with human endeavours.[91] This can include using AI to engage in academic plagiarism (cheating) or violations of others' copyrights and intellectual property.[92]

Proponents of humanistic endeavours, from artists to sportspeople, may have some reason to be concerned. After being beaten by Google *DeepMind's AlphaGo*, Lee Sedol retired from the board game Go. Lee, a top-rated master of the ancient Chinese game, was quoted as saying:

> With the debut of AI in Go games, I've realized that I'm not at the top even if I become the number one through frantic efforts .... Even if I become the number one, there is an entity [i.e., AI] that cannot be defeated ....[93]

*Examples in Chapter 2: Harms of video games/multimedia* and *social media* (vis-à-vis addiction), IBM's *Watson* and Google's *AlphaGo* defeating human champions (in Jeopardy! and Go respectively), *ChatGPT* leading the way for Generative AI.

*Considerations to be aware of*:

- Being aware of the time one spends with technology, the degree one is dependent on its use, and its *potential harms to genuine human connection.*
- Setting boundaries with technology, such as incorporating *digital detox* and *mindful* practices.
- Understanding the risks involved in using technologies such as Generative AI, which can lead to *violations of copyright and academic integrity.*

- Distinguishing between *instrumental* and *intrinsic* activities, and why technology should not be used in defeating the purpose of humanistic endeavours.

* * *

As with the timeline in Chapter 2, this taxonomy of issues in digital ethics is not exhaustive. In fact, the taxonomy serves as a building block for a deeper understanding (and future research) on up-and-coming problems. We may not know what is in store for technology and society in the years ahead, but we hope you now have an initial sense of how we might categorise new issues on the horizon. Hopefully, you now have some idea about the ways in which technology can impact yourself, the people around you, and the wider world.

Knowing the issues is one thing; but analysing and interrogating them, in the process of trying to gain a deeper understanding, is another. In Chapter 1, we briefly introduced you to philosophy. In the next chapter, we will begin to hone your philosophical skills, with a concise (and gentle) introduction to ethical theories.

## Notes

1. A wide range of literature is part of this discussion, including: Floridi (2023), Müller (2023), Paltiel, Cheong, Coghlan, & Lederman (2023), and Ethics of Artificial Intelligence (n.d.).
2. As an interesting experiment, we queried all three of OpenAI's GPT4 (gpt-4o), Anthropic's Claude 3.5 Sonnet, and Google's Gemini 2.0 Flash (versions as of 10 April 2025). The common prompt we issued to all these LLMs was: 'classify the issues in digital ethics'; the top-level headers are then considered in our categorisation process.
3. From the Oxford Dictionary of English, as reproduced in the Dictionary app in MacOS.
4. Mason (1986).
5. McCandless, Evans, & Barton (2024).
6. Inspired by tips from organisations and government bodies: 10 TIPS To protect Your Privacy On the Internet. (n.d.), Surveillance Self-Defense (n.d.), and OAIC (2023b, March 10).
7. Have I Been Pwned (n.d.).
8. Mozilla Monitor (n.d.).
9. REGULATION (EU) 2016/679 OF THE EUROPEAN PARLIAMENT AND OF THE COUNCIL of 27 April 2016 on the Protection of Natural Persons with Regard to the Processing of Personal Data and on the Free Movement of Such Data, and Repealing Directive 95/46/EC (General Data Protection Regulation), (EU) 2016/679 (2016). https://gdpr-info.eu/

   Hereinafter, this will be referred to as Reg 2016/679 (2016), for brevity.

10. Floridi (2018).
11. Mason (1986).
12. Petzold, Roggendorf, Rowshankish, & Sporleder (2020, June 26).
13. Ensemble definition based on: FAQ (n.d.) and Ladley (2012).
14. GDPR compliance checklist (n.d.).
15. Reg 2016/679 (2016).
16. Leisterer (2015, February 16).
17. OAIC (2023a, March 10).
18. 2014 Schedule (n.d.).
19. For a complete text on the technical considerations on fairness, we recommend strongly the FAIR ML book by Barocas, Hardt, & Narayanan (2023).
20. Barocas, Hardt, & Narayanan (2023).
21. Chouldechova (2016, October 24).
22. Narayanan (Director). (2018, March 2).
23. Venkatasubramanian & Alfano (2020).
24. Reg 2016/679 (2016).
25. Examples are in Rudin (2019) and Shah (2018).
26. Murray, Cheong, & Paterson (2023, July 10).
27. Elish (2019).
28. Sloan (2023, May 31).
29. Jacovi, Marasović, Miller, & Goldberg (2021).
30. Susnjara & Smalley (2021, July 29).
31. We will consider this trade-off in Chapter 6.
32. Miller (2019).
33. Rudin (2019).
34. Miller (2019).
35. Lorenz-Spreen et al. (2021).
36. To borrow a commonly used definition, as cited, from the Australian Human Rights Commission: Classroom resources on equality and equity - Australian Human Rights Commission. (n.d.).
37. Johnson (2019, October 24).
38. Njoto, Cheong, Frermann, & Ruppanner (2024) and Njoto et al. (2022).
39. Asaro (2019).
40. Alfano et al. (2024) and Cheong et al. (2024).
41. OECD. (2000).
42. Acosta Rueda, Rouse, Cheong, & Ruppanner (2022) and van Deursen & Helsper (2015).
43. Sheldrick et al. (2025, January 7) and Sheldrick, Chang, Kurnia, & McKay (2023).
44. A resource website on the proposed EU AI Act is maintained by the Future of Life Institute at: EU Artificial Intelligence Act | Up-to-date developments and analyses of the EU AI Act. (n.d.).
45. An example includes the planned shutdown of the 3G network in Australia potentially affecting owners of existing (more modern) 4G devices as well, due to a quirk in compatibility. See coverage in: Hugh Jeffreys (Director). (2023, November 26) and Williams (2024, March 11).
46. A key textbook in HCI is by Sharp, Rogers, & Preece (2023).
47. The Facebook Files (2021, October 1).
48. Osman (2025).
49. Press coverage includes Howard (2023, June 7) and Yang (2024, October 24).

50. Bozdag (2013).
51. Nguyen (2020).
52. Quintana et al. (2022) and World Health Organization et al. (2020, September 23).
53. Ward (2018).
54. Alfano, Carter, & Cheong (2018).
55. Coghlan et al. (2023)
56. Anguiano (2025, January 6).
57. Examples of these can be found in Shaban (2025, January 20) and The Associated Press (2024, October 19).
58. AP (1983, August 11).
59. AP/Reuters (2024, August 7).
60. Statistics are based on Data centres & networks. (n.d.) and Ramachandran, Stewart, Hardin, & Crossan (2024, November 19).
61. Statistics are based on: DOE Releases New Report Evaluating Increase in Electricity Demand from Data Centers. (2024, December 20) and Yang (2024, July 25).
62. Patterson et al. (2021).
63. Rough calculations using a tool by US EPA (2015, August 28).
64. Luccioni, Jernite, & Strubell (2024).
65. Dastin & Nellis (2023, February 23).
66. Balaram (2019).
67. Wang, Yang, Wang, & Huang (2024).
68. Bulow. (1986).
69. Chen (2021, July 14) and Sabbaghi, Cade, Behdad, & Bisantz (2017).
70. Kainz & Broaddus (2025, April 22).
71. Right to repair. (2021).
72. Klosowski (2021, July 15) and Ozturkcan (2024).
73. Chen (2021, July 14).
74. NRC (2022).
75. Coghlan & Parker (2023).
76. Examples can be found in: Morrison (2025, May 3) and West (2024, November 7).
77. For further context, please see sources such as The Smithsonian Magazine: Conniff (2011).
78. What is the gig economy and what's the deal for gig workers? (2024, November 22).
79. Berger, Chen, & Frey (2018).
80. A multi-disciplinary discussion can be found in Möhlmann, Zalmanson, Henfridsson, & Gregory (2021) and Wang & Churchill (2025).
81. See Ruppanner et al. (2023), which cites earlier statistics in O'Grady, Higgins, Giron, & Kearney (2022, August 8).
82. Woodruff et al. (2024).
83. Global Views on AI 2023 (2023).
84. Hendricks (2020, June 10).
85. Stupp (2019, August 30).
86. Interestingly, as educators, the authors and other lecturers seek to increase 'active learning' quizzes using smartphones, rather than getting distracted by the various memes and TikTok videos.
87. Bhattacharya, Bashar, Srivastava, & Singh (2019).
88. Luker (2022).
89. Grissinger (2019).
90. Hirose & Olson (Eds.). (2015) and Weber, Roth, & Wittich (1978).

91. Covered in the press: Geraets (2022, November 11) and Roose (2022, September 2).
92. Marcus & Southen (2024, January 6) and Van Noorden & Perkel (2023).
93. Cheong-mo (2019, November 27).

## References


10 TIPS To protect Your Privacy On the Internet. (n.d.). European Parliament. Retrieved June 12, 2025, from https://www.europarl.europa.eu/ep_products/dataprotection/index_en.html

2014 Schedule. (n.d.). FAT ML. Retrieved June 12, 2025, from https://www.fatml.org/schedule/2014

Acosta Rueda, L., Rouse, L. A., Cheong, M., & Ruppanner, L. (2022). Mapping the Digital Divide in Victoria. The Future of Work Lab.

Alfano, M., Abedin, E., Reimann, R., Ferreira, M., & Cheong, M. (2024). Now you see me, now you don't: An exploration of religious exnomination in DALL-E. Ethics and Information Technology, 26(2), 27. https://doi.org/10.1007/s10676-024-09760-y

Alfano, M., Carter, J. A., & Cheong, M. (2018). Technological seduction and self-radicalization. Journal of the American Philosophical Association, 4(3), 298–322. https://doi.org/10.1017/apa.2018.27

Anguiano, D. (2025, January 6). LA Tech Entrepreneur Nearly Misses Flight After Getting Trapped in Robotaxi. The Guardian. https://www.theguardian.com/us-news/2025/jan/06/man-trapped-waymo-los-angeles

AP. (1983, August 11). AROUND THE NATION; Jury Awards $10 Million in Killing by Robot. New York Times, Late Edition (East Coast), A.12.

AP/Reuters. (2024, August 7). "Garbage" Safety Culture at Boeing to Blame for Cabin Door Blowout Mid-flight, Watchdog Hears. ABC News. https://www.abc.net.au/news/2024-08-07/nobodys-accountable-investigation-boeing-plane-door-737-max/104194554

Arvind Narayanan (Director). (2018, March 2). Tutorial: 21 Fairness Definitions and Their Politics [Video recording]. https://www.youtube.com/watch?v=jIXIuYdnyyk

Asaro, P. M. (2019). AI ethics in predictive policing: From models of threat to an ethics of care. IEEE Technology and Society Magazine, 38(2), 40–53. https://doi.org/10.1109/MTS.2019.2915154

Balaram, V. (2019). Rare earth elements: A review of applications, occurrence, exploration, analysis, recycling, and environmental impact. Geoscience Frontiers, 10(4), 1285–1303. https://doi.org/10.1016/j.gsf.2018.12.005

Barocas, S., Hardt, M., & Narayanan, A. (2023). Fairness and Machine Learning: Limitations and Opportunities. The MIT Press.

Berger, T., Chen, C., & Frey, C. B. (2018). Drivers of disruption? Estimating the uber effect. European Economic Review, 110, 197–210. https://doi.org/10.1016/j.euroecorev.2018.05.006

Bhattacharya, S., Bashar, M. A., Srivastava, A., & Singh, A. (2019). NOMOPHOBIA: NO MObile PHone PhoBIA. Journal of Family Medicine and Primary Care, 8(4), 1297. https://doi.org/10.4103/jfmpc.jfmpc_71_19

Bozdag, E. (2013). Bias in algorithmic filtering and personalization. Ethics and Information Technology, 15(3), 209–227. https://doi.org/10.1007/s10676-013-9321-6

Bulow, J. (1986). An economic theory of planned obsolescence. The Quarterly Journal of Economics, 101(4), 729–749. https://doi.org/10.2307/1884176

Chen, B. X. (2021, July 14). Why You Should Care About Your Right to Repair Gadgets. The New York Times. https://www.nytimes.com/2021/07/14/technology/personaltech/right-to-repair-iphones-android.html

Cheong, M., Abedin, E., Ferreira, M., Reimann, R., Chalson, S., Robinson, P., Byrne, J., Ruppanner, L., Alfano, M., & Klein, C. (2024). Investigating gender and racial biases in DALL-E Mini images. ACM Journal on Responsible Computing, 1(2), 13:1–13:20. https://doi.org/10.1145/3649883

Cheong-mo, Y. (2019, November 27). (Yonhap Interview) Go Master Lee Says He Quits Unable to Win Over AI Go Players. Yonhap News Agency. https://en.yna.co.kr/view/AEN20191127004800315

Chouldechova, A. (2016, October 24). Fair Prediction with Disparate Impact: A Study of Bias in Recidivism Prediction Instruments. arXiv.Org. https://arxiv.org/abs/1610.07524v1

Classroom resources on equality and equity – Australian Human Rights Commission. (n.d.). Australian Human Rights Commission. Retrieved June 12, 2025, from https://humanrights.gov.au/classroom-resources-equality-and-equity

Coghlan, S., Leins, K., Sheldrick, S., Cheong, M., Gooding, P., & D'Alfonso, S. (2023). To chat or bot to chat: Ethical issues with using chatbots in mental health. DIGITAL HEALTH, 9, 20552076231183542. https://doi.org/10.1177/20552076231183542

Coghlan, S., & Parker, C. (2023). Harm to nonhuman animals from AI: A systematic account and framework. Philosophy & Technology, 36(2), 25. https://doi.org/10.1007/s13347-023-00627-6

Conniff, R. (2011). What the Luddites Really Fought Against. Smithsonian Magazine. https://www.smithsonianmag.com/history/what-the-luddites-really-fought-against-264412/

Dastin, J., & Nellis, S. (2023, February 23). Focus: For Tech Giants, AI Like Bing and Bard Poses Billion-dollar Search Problem. Reuters. https://www.reuters.com/technology/tech-giants-ai-like-bing-bard-poses-billion-dollar-search-problem-2023-02-22/

Data centres & networks. (n.d.). IEA. Retrieved June 14, 2025, from https://www.iea.org/energy-system/buildings/data-centres-and-data-transmission-networks

DOE Releases New Report Evaluating Increase in Electricity Demand from Data Centers. (2024, December 20). Energy.Gov. https://www.energy.gov/articles/doe-releases-new-report-evaluating-increase-electricity-demand-data-centers

Elish, M. C. (2019). Moral crumple zones: Cautionary tales in human-robot interaction. Engaging Science, Technology, and Society, 5, 40–60. https://doi.org/10.17351/ests2019.260

Ethics of Artificial Intelligence. (n.d.). UNESCO. Retrieved June 12, 2025, from https://www.unesco.org/en/artificial-intelligence/recommendation-ethics

EU Artificial Intelligence Act | Up-to-date developments and analyses of the EU AI Act. (n.d.). Future of Life Institute. Retrieved June 12, 2025, from https://artificialintelligenceact.eu/

FAQ. (n.d.). [Digital Trade & Data Governance Hub]. Retrieved June 12, 2025, from https://datagovhub.elliott.gwu.edu/faq/

Floridi, L. (2018). Soft ethics, the governance of the digital and the general data protection regulation. Philosophical Transactions of the Royal Society A: Mathematical, Physical and Engineering Sciences, 376(2133), 20180081. https://doi.org/10.1098/rsta.2018.0081

Floridi, L. (2023). The Ethics of Artificial Intelligence: Principles, Challenges, and Opportunities. Oxford University Press. https://doi.org/10.1093/oso/9780198883098.001.0001

GDPR compliance checklist. (n.d.). GDPR.Eu. Retrieved June 12, 2025, from https://gdpr.eu/checklist/

Geraets, N. (2022, November 11). Dress Code: Does AI Technology Belong in Fashion? The Sydney Morning Herald. https://www.smh.com.au/business/companies/dress-code-does-ai-technology-belong-in-fashion-20221108-p5bwh2.html

Global Views on AI 2023. (2023). Ipos. https://www.ipsos.com/sites/default/files/ct/news/documents/2023-07/Ipsos%20Global%20AI%202023%20Report-WEB.pdf

Grissinger, M. (2019). Understanding human over-Reliance on technology. Pharmacy and Therapeutics, 44(6), 320–375.

Have I Been Pwned. (n.d.). Have I Been Pwned. Retrieved June 12, 2025, from https://haveibeenpwned.com/

Hendricks, K. (2020, June 10). Investigating Scam Phone Calls. FBI: Law Enforcement Bulletin. https://leb.fbi.gov/articles/featured-articles/investigating-scam-phone-calls

Hirose, I., & Olson, J. (Eds.). (2015). The Oxford Handbook of Value Theory. Oxford Univ. Press. https://doi.org/10.1093/oxfordhb/9780199959303.001.0001

Howard, J. (2023, June 7). ChatGPT's Responses to Suicide, Addiction, Sexual Assault Crises Raise Questions in New Study. CNN. https://www.cnn.com/2023/06/07/health/chatgpt-health-crisis-responses-wellness

IISC, & Maguire, A. (2016, January 14). Illustrating Equality VS Equity. Interaction Institute for Social Change. https://interactioninstitute.org/illustrating-equality-vs-equity/

Jacovi, A., Marasović, A., Miller, T., & Goldberg, Y. (2021). Formalizing trust in artificial intelligence: Prerequisites, causes and goals of human trust in AI. Proceedings of the 2021 ACM Conference on Fairness, Accountability, and Transparency, 624–635. https://doi.org/10.1145/3442188.3445923

Jeffreys, H. (Director). (2023, November 26). Carriers are Killing 4G & 3G Devices – Your 4G Phone May Soon Stop Working [Video recording]. https://www.youtube.com/watch?v=Q6qb9dml6So

Johnson, C. Y. (2019, October 24). Racial Bias in a Medical Algorithm Favors White Patients Over Sicker Black Patients. The Washington Post. https://www.washingtonpost.com/health/2019/10/24/racial-bias-medical-algorithm-favors-white-patients-over-sicker-black-patients/

Kainz, N., & Broaddus, A. (2025, April 22). Right-to-repair Revolution: Farmers Challenge John Deere's Control Over Equipment Repair. NBC News. https://www.nbcnews.com/business/consumer/right-to-repair-farmers-challenge-john-deere-control-equipment-rcna199651

Klosowski, T. (2021, July 15). What You Should Know About Right to Repair. Wirecutter. https://www.nytimes.com/wirecutter/blog/what-is-right-to-repair/

Ladley, J. (2012). Data Governance: How to Design, Deploy, and Sustain an Effective Data Governance Program. Morgan Kaufmann.

Leisterer, H. (2015, February 16). The Principle of 'Datensparsamkeit.' Digital Society Blog. https://www.hiig.de/en/the-principle-of-datensparsamkeit/

Lorenz-Spreen, P., Geers, M., Pachur, T., Hertwig, R., Lewandowsky, S., & Herzog, S. M. (2021). Boosting people's ability to detect microtargeted advertising. Scientific Reports, 11(1), 15541. https://doi.org/10.1038/s41598-021-94796-z

Luccioni, A. S., Jernite, Y., & Strubell, E. (2024). Power hungry processing: Watts driving the cost of AI deployment? The 2024 ACM Conference on Fairness, Accountability, and Transparency, 85–99. https://doi.org/10.1145/3630106.3658542

Luker, E. (2022). Video Games and Screen Addiction. Mayo Clinic Health System. https://www.mayoclinichealthsystem.org/hometown-health/speaking-of-health/are-video-games-and-screens-another-addiction

Marcus, G., & Southen, R. (2024, January 6). Generative AI has a Visual Plagiarism Problem – IEEE Spectrum. https://spectrum.ieee.org/midjourney-copyright

Mason, R. O. (1986). Four ethical issues of the information age. MIS Quarterly, 10(1), 5–12. https://doi.org/10.2307/248873

McCandless, D., Evans, T., & Barton, P. (2024). World's Biggest Data Breaches & Hacks. Information is Beautiful. https://informationisbeautiful.net

Miller, T. (2019). Explanation in artificial intelligence: Insights from the social sciences. Artificial Intelligence, 267, 1–38. https://doi.org/10.1016/j.artint.2018.07.007

Möhlmann, M., Zalmanson, L., Henfridsson, O., & Gregory, R. W. (2021). Algorithmic management of work on online labor platforms: When matching meets control. MIS Quarterly, 45(4), 1999–2022.

Morrison, C. (2025, May 3). Majority of Canadians Say Misinformation Affected Federal Election: Poll—National. The Canadian Press. https://globalnews.ca/news/11162689/canadian-misinformation-federal-election-poll/

Mozilla Monitor. (n.d.). Mozilla Monitor. Retrieved June 12, 2025, from https://monitor.mozilla.org

Müller, V. C. (2023). Ethics of artificial intelligence and robotics. In E. N. Zalta, & U. Nodelman (Eds.), The Stanford Encyclopedia of Philosophy (Fall 2023). Metaphysics Research Lab, Stanford University. https://plato.stanford.edu/archives/fall2023/entries/ethics-ai/

Murray, T., Cheong, M., & Paterson, J. (2023, July 10). The Flawed Algorithm at the Heart of Robodebt. The Pursuit. https://pursuit.unimelb.edu.au/articles/the-flawed-algorithm-at-the-heart-of-robodebt

Nguyen, C. T. (2020). ECHO CHAMBERS AND EPISTEMIC BUBBLES. Episteme, 17(2), 141–161. https://doi.org/10.1017/epi.2018.32

Njoto, S., Cheong, M., Frermann, L., & Ruppanner, L. (2024). Bias and discrimination against women and parents in semi-automated hiring systems. New Technology, Work and Employment, 1–11. https://doi.org/10.1111/ntwe.12321

Njoto, S., Cheong, M., Lederman, R., McLoughney, A., Ruppanner, L., & Wirth, A. (2022). Gender bias in AI recruitment systems: A sociological-and data science-based case study. 2022 IEEE International Symposium on Technology and Society (ISTAS), 1, 1–7. https://doi.org/10.1109/ISTAS55053.2022.10227106

NRC. (2022). Backgrounder on the Three Mile Island Accident. NRC. https://www.nrc.gov/reading-rm/doc-collections/fact-sheets/3mile-isle.html

O'Grady, M., Higgins, S., Giron, F., & Kearney, M. (2022, August 8). Future of Jobs Forecast, 2020 to 2040 (India, China,… Forrester. https://www.forrester.com/report/future-of-jobs-forecast-2020-to-2040-india-china-south-korea-australia-and-japan/RES177789

OAIC. (2023a, March 10). Australian Privacy Principles. OAIC. https://www.oaic.gov.au/privacy/australian-privacy-principles

OAIC. (2023b, March 10). Tips to Protect Your Privacy. OAIC. https://www.oaic.gov.au/privacy/your-privacy-rights/ways-to-protect-your-privacy/tips-to-protect-your-privacy

OECD. (2000). Understanding the Digital Divide (OECD Digital Economy Papers). OECD Publishing. https://www.oecd.org/en/publications/understanding-the-digital-divide_236405667766.html

Osman, M. (2025). Psychological harm: What is it and how does it apply to consumer products with internet connectivity? Journal of Risk Research, 1–22. https://doi.org/10.1080/13669877.2025.2491086

Ozturkcan, S. (2024). The right-to-repair movement: Sustainability and consumer rights. Journal of Information Technology Teaching Cases, 14(2), 217–223. https://doi.org/10.1177/20438869231178037

Paltiel, M., Cheong, M., Coghlan, S., & Lederman, R. (2023). Approaches and models for teaching digital ethics in information systems courses – A review of the literature. Australasian Journal of Information Systems, 27. https://doi.org/10.3127/ajis.v27i0.4517

Patterson, D., Gonzalez, J., Le, Q., Liang, C., Munguia, L.-M., Rothchild, D., So, D., Texier, M., & Dean, J. (2021). Carbon Emissions and Large Neural Network Training (No. arXiv:2104.10350). arXiv. https://doi.org/10.48550/arXiv.2104.10350

Petzold, B., Roggendorf, M., Rowshankish, K., & Sporleder, C. (2020, June 26). Designing Data Governance that Delivers Value | McKinsey. McKinsey Digital. https://www.mckinsey.com/capabilities/mckinsey-digital/our-insights/designing-data-governance-that-delivers-value

Quintana, I. O., Reimann, R., Cheong, M., Alfano, M., & Klein, C. (2022). Polarization and trust in the evolution of vaccine discourse on Twitter during COVID-19. PLOS ONE, 17(12), e0277292. https://doi.org/10.1371/journal.pone.0277292

Ramachandran, K., Stewart, D., Hardin, K., & Crossan, G. (2024, November 19). As Generative AI Asks for More Power, Data Centers Seek More Reliable, Cleaner Energy Solutions. Deloitte Insights. https://www2.deloitte.com/us/en/insights/industry/technology/technology-media-and-telecom-predictions/2025/genai-power-consumption-creates-need-for-more-sustainable-data-centers.html

REGULATION (EU) 2016/679 OF THE EUROPEAN PARLIAMENT AND OF THE COUNCIL of 27 April 2016 on the Protection of Natural Persons with Regard to the Processing of Personal Data and on the Free Movement of Such Data, and Repealing Directive 95/46/EC (General Data Protection Regulation), (EU) 2016/679 (2016). https://gdpr-info.eu/

Right to repair. (2021). Productivity Commission. https://www.pc.gov.au/inquiries/completed/repair/report

Roose, K. (2022, September 2). An A.I.-Generated Picture Won an Art Prize. Artists Aren't Happy. The New York Times. https://www.nytimes.com/2022/09/02/technology/ai-artificial-intelligence-artists.html

Rudin, C. (2019). Stop Explaining Black Box Machine Learning Models for High Stakes Decisions and Use Interpretable Models Instead (No. arXiv:1811.10154). arXiv. https://doi.org/10.48550/arXiv.1811.10154

Ruppanner, L., Churchill, B., Bissell, D., Ghin, P., Hydelund, C., Ainsworth, S., Blackham, A., Borland, J., Cheong, M., Evans, M., Frermann, L., King, T., & Vetere, F. (2023). 2023 State of the Future of Work (Australia) [Report]. University of Melbourne. https://apo.org.au/node/322034

Sabbaghi, M., Cade, W., Behdad, S., & Bisantz, A. M. (2017). The current status of the consumer electronics repair industry in the U.S.: A survey-based study. Resources, Conservation and Recycling, 116, 137–151. https://doi.org/10.1016/j.resconrec.2016.09.013

Shaban, B. (2025, January 20). While Waymo Not Blamed in Multi-car Wreck, It's the First Fatal Collision Involving a Driverless Car. NBC Bay Area. https://www.nbcbayarea.com/investigations/waymo-multi-car-wreck-san-francisco-driverless/3766860/

Shah, H. (2018). Algorithmic accountability. Philosophical Transactions of the Royal Society A: Mathematical, Physical and Engineering Sciences, 376(2128), 20170362. https://doi.org/10.1098/rsta.2017.0362

Sharp, H., Rogers, Y., & Preece, J. (2023). Interaction Design: Beyond Human-Computer Interaction (6th ed.). John Wiley & Sons, Inc.

Sheldrick, S., Andersson, J., Chang, S., Mckay, D., & Kurnia, S. (2025, January 7). Hooja? How is AI impacting the digital divide for Norrland's SMEs. Proceedings of the 58th Hawaii International Conference on System Sciences. HICSS, Hawaii. https://hdl.handle.net/10125/109649

Sheldrick, S., Chang, S., Kurnia, S., & McKay, D. (2023). Across the Great Digital Divide: Investigating the Impact of AI on Rural SMEs. PACIS 2023 Proceedings, 17 pages. https://aisel.aisnet.org/pacis2023/13

Sloan, K. (2023, May 31). A Lawyer Used ChatGPT to Cite Bogus Cases. What are the Ethics? Reuters. https://www.reuters.com/legal/transactional/lawyer-used-chatgpt-cite-bogus-cases-what-are-ethics-2023-05-30/

Stupp, C. (2019, August 30). Fraudsters Used AI to Mimic CEO's Voice in Unusual Cybercrime Case. Wall Street Journal. https://www.wsj.com/articles/fraudsters-use-ai-to-mimic-ceos-voice-in-unusual-cybercrime-case-11567157402

Surveillance Self-Defense - Basics. (n.d.). EFF. Retrieved June 12, 2025, from https://ssd.eff.org/module-categories/basics

Susnjara, S., & Smalley, I. (2021, July 29). What Is Open Source Software? IBM. https://www.ibm.com/think/topics/open-source

The Associated Press. (2024, October 19). U.S. to Probe Tesla's "Full Self-Driving" System After Pedestrian Killed. NPR. https://www.npr.org/2024/10/19/g-s1-29030/us-probe-tesla-full-self-driving-system

The Facebook Files. (2021, October 1). Wall Street Journal. https://www.wsj.com/articles/the-facebook-files-11631713039

US EPA. (2015, August 28). Greenhouse Gas Equivalencies Calculator [Data and Tools]. https://www.epa.gov/energy/greenhouse-gas-equivalencies-calculator

van Deursen, A. J. A. M., & Helsper, E. J. (2015). The third-level digital divide: Who benefits most from being online? In L. Robinson, S. R. Cotten, J. Schulz, T. M. Hale, & A. Williams (Eds.), Studies in Media and Communications (Vol. 10, pp. 29–52). Emerald Group Publishing Limited. https://doi.org/10.1108/S2050-206020150000010002

Van Noorden, R., & Perkel, J. M. (2023). AI and science: What 1,600 researchers think. Nature, 621(7980), 672–675. https://doi.org/10.1038/d41586-023-02980-0

Venkatasubramanian, S., & Alfano, M. (2020). The philosophical basis of algorithmic recourse. Proceedings of the 2020 Conference on Fairness, Accountability, and Transparency, 284–293. https://doi.org/10.1145/3351095.3372876

Wang, Q., & Churchill, B. (2025). Risky business: How food-delivery platform riders understand and manage safety at work. Journal of Sociology, 61(2), 291–307. https://doi.org/10.1177/14407833241246571

Wang, W., Yang, Y., Wang, D., & Huang, L. (2024). Toxic effects of rare earth elements on human health: A review. Toxics, 12(5), 317. https://doi.org/10.3390/toxics12050317

Ward, K. (2018). Social networks, the 2016 US presidential election, and kantian ethics: Applying the categorical imperative to Cambridge Analytica's behavioral microtargeting. Journal of Media Ethics, 33(3), 133–148. https://doi.org/10.1080/23736992.2018.1477047

Weber, M., Roth, G., & Wittich, C. (1978). Economy and Society: An Outline of Interpretative Sociology. University of California press.

West, D. M. (2024, November 7). How Disinformation Defined the 2024 Election Narrative. Brookings. https://www.brookings.edu/articles/how-disinformation-defined-the-2024-election-narrative/

What is the gig economy and what's the deal for gig workers? (2024, November 22). World Economic Forum. https://www.weforum.org/stories/2024/11/what-gig-economy-workers/

Williams, T. (2024, March 11). Your 4G Phone Could be Impacted by 3G Shutdowns this Year. Here's What to Know. ABC News. https://www.abc.net.au/news/2024-03-12/4g-phones-volte-regional-customers-australia-3g-network-shutdown/103494136

Woodruff, A., Shelby, R., Kelley, P. G., Rousso-Schindler, S., Smith-Loud, J., & Wilcox, L. (2024). How knowledge workers think generative AI will (not) transform their industries. Proceedings of the 2024 CHI Conference on Human Factors in Computing Systems, 1–26. https://doi.org/10.1145/3613904.3642700

World Health Organization, United Nations, UNICEF, UNDP, UNESCO, UNAIDS, ITU, UN Global Pulse, & IFRC. (2020, September 23). Managing the COVID-19 Infodemic: Promoting Healthy Behaviours and Mitigating the Harm from Misinformation and Disinformation. World Health Organization. https://www.who.int/news/item/23-09-2020-managing-the-covid-19-infodemic-promoting-healthy-behaviours-and-mitigating-the-harm-from-misinformation-and-disinformation

Yang, A. (2024, October 24). Lawsuit Claims Character.AI is Responsible for Teen's Suicide. NBC News. https://www.nbcnews.com/tech/characterai-lawsuit-florida-teen-death-rcna176791

Yang, S. (2024, July 25). Power-hungry Data Centres Scrambling to Find Enough Electricity to Meet Demand. ABC News. https://www.abc.net.au/news/2024-07-26/data-centre-electricity-grid-demand/104140808

# 4

# Philosophical and Ethical Theory

## Introduction to Moral Theory

The last few chapters have introduced us to a birds-eye overview of how ethicists think, and the various issues under our umbrella of digital ethics. To provide you with a concrete foundation in ethical thought, this chapter will explain the four most prominent ethical or moral theories.

Recall in Chapter 1 that there are different philosophical approaches to ethics *itself* (metaethics). We have said that ethics is not simply a matter of gut feeling or taking surveys of what most people think about an issue. Instead, ethics can require giving reasons and argument to justify a claim that something, such as some controversial use of a digital technology, has a particular ethical status (right vs wrong, justified vs unjustified, and so forth).

Now, we look at the various *theories* one can turn to for assistance when considering whether something is ethical or otherwise.[1] Four popular theories are **utilitarianism**, **deontology**, **virtue ethics**, and **care ethics**. These theories tend to highlight different aspects of morality. Utilitarianism highlights consequences of actions, deontology highlights duties or obligations, virtue ethics highlights our moral characters, and care ethics highlights caring relations.

Now, a word of advice: It is true that every moral theory can be controversial. It is also true these theories come from only one part of human thought, namely Western philosophy. However, we think it is worth knowing about these theories, partly because they are widely referred to, and discussed in, digital ethics. Before we discuss them, a bit of background about moral theory will help us to understand the point of these theories and why they have achieved a high profile in moral philosophy, and in digital ethics.

Again, recall that in Chapter 1 that we looked at one aspect of the ethics of social media platforms involving Facebook, which we reproduce here[2]:

> Facebook allowed teenagers to be exposed to content about unrealistic body image that made them mentally unwell.
>
> Therefore, Facebook has done the wrong thing to some teenagers as a result of this.

Due to the socially troubling effects of social media, some governments have moved to regulate them. This includes, in some places, banning children

DOI: 10.1201/9781003393511-4

below a certain age from accessing social media platforms.[3] Is that the right way to go? Actually, it is quite controversial, and opinions widely differ.

Now consider some other tricky or at least contested questions that digital ethicists confront:

1. Is it ever acceptable to roll out modern surveillance technology in public places to catch criminals or find missing children? If so, under what conditions is that permissible?
2. Is it ethically justified to decide on welfare payments or loans using algorithms that are 'better than humans' but which operate as 'black boxes' (whose inner workings and calculations are unclear to us) and may be occasionally biased?
3. Ought we to use advanced artificial intelligence (AI) systems to fully automate at least some kinds of medical decision-making, or should we only ever use AI to provide decision-support to doctors and nurses?
4. Should governments ban and stop working on automated drones or robots that can track and kill the enemy fully independently of human involvement? What if other countries are vigorously working on such weapons?
5. Should we explore human enhancement technologies that could, for example, merge with our brains to boost intelligence or make us more compassionate or cooperative?

As you may agree after reading these questions, sometimes the answers to digital ethical issues are not always clear cut. But even if we personally think the answers are perfectly clear to some of these questions, there will often be other thoughtful people who disagree with us. We may also want to keep an open mind on some issues until we learn more about the facts and have had time to carefully weigh the ethical pros and cons. Ethical questions can be both difficult and controversial. Uncertainty and confusion are common, and we do best to approach moral questions with a degree of humility, avoiding arrogance and judgementalism. One might know a lot about a given technology but still be mistaken about whether its use is ethical.

Occasionally, we may even encounter a moral dilemma – a situation where all the apparent options are or seem wrong – that leaves us genuinely troubled or at least uncertain. A quintessential philosophical example of a moral dilemma is the famous **Trolley Problem**.[4] It looks something like this:

> You are standing next to a switch controlling a set of train (trolley) tracks.
>
> You see a train hurtling out of control, about to collide with five people stuck on track A.
>
> There is, however, a sidetrack (track B), where a single person is crossing.

> You could do nothing, and let the train continue on track A, where it will kill the five people. Or, you could activate the switch to divert the out-of-control train onto track B. This would save the five people but kill the one person. Should you pull the switch?

This kind of example has been used in digital ethics, especially in the context of self-driving cars.[5] There the moral dilemma is said to concern how we should program automated vehicles for hypothetical occasions where they must 'decide' who or how many people to run into. Imagine an older person unexpectedly walks onto the road and the car, due to its high speed, must either run into the older person or crash into a tree and kill the young driver. What should the smart car do?

Such examples may seem somewhat artificial and unrealistic. However, there are plenty of real-life cases where all options may seem morally dubious or where we are left puzzled and troubled about what to think. In making a well-supported and thoughtful ethical decision or judgement, it will often be important to understand certain facts, such as facts about the technology and about the way people and society interact with it. It will also often be important to think through the ethical ideas and arguments that we are using. This will include considering counterarguments to a position or judgement, as we have alluded to earlier as well. Identifying and addressing possible weaknesses in an argument helps to make the argument stronger. A well-defended ethical argument on a complex and contestable issue about digital technology thus requires considerable thought and understanding.

Of course, on some occasions, it will indeed be obvious or at least relatively clear what is morally right and wrong. In 2016, Microsoft's supposedly cutting-edge chatbot 'Tay' – an acronym for '*Thinking About You*' – learnt from users how to produce horribly racist, sexist, Holocaust-denying tweets.[6] The public chatbot was swiftly shut down by the company, which posted that it was 'deeply sorry for the unintended offensive and hurtful tweets from Tay'.[7] Quite obviously, letting Tay continue to run rampant would have been wrong, as Microsoft knew.

Critics also argued that Tay should never have been publicly released in the first place. Despite Microsoft doing some pre-testing of the chatbot's safety, it should have been obvious, critics argued, that internet trolls would find ways to make it tweet hateful messages. Critics alleged that Microsoft failed to think through the consequences of putting the chatbot online and thus failed to show sufficient responsibility, even though they did exhibit some accountability after the incident had occurred.

Another digital technology incident that received widespread condemnation was the so-called *Facebook–Cambridge Analytica* data scandal (outlined in Chapters 2 and 3). Facebook had allowed the political consultancy group Cambridge Analytica to harvest personal data without consent or disclosure from the accounts of millions of Facebook users.[8] Cambridge Analytica then used the data to psychologically profile users to try to influence the 2016

US presidential election. As in the Microsoft Tay case, Facebook later apologised for wrongdoing, with Mark Zuckerberg saying that his company had breached the trust of its customers.

Nonetheless, in many other cases involving digital technologies, it will be much harder to know what the right or wrong action is. And even in cases where it is clear what we should or shouldn't do, understanding exactly *why* it is right or wrong, and making sure those reasons are consistent with our other moral beliefs, can be more difficult. In view of the difficulty of moral judgement and argument, philosophers have sought over decades and centuries to provide more systematic approaches to moral questions. The construction of ethical theories is the most high-profile result of these attempts. Each normative theory is a product of dedicated thought and long argument regarding questions of how to act, respond, behave, and live. Moral theorists try to go to the heart of what makes actions or behaviour right or wrong.

It is important to appreciate that ethical theories are highly *general*. For example, as we shall see, virtue ethics tells us to follow the example of highly virtuous or good and morally wise people. Utilitarianism instructs us to produce the best consequences that we can. This is pretty general advice. Thus, virtue ethics and utilitarianism, and the other moral theories as well, do not directly or immediately answer specific and contextual moral questions but rather give broad guidance on how to answer them.

Yet the somewhat abstract nature of theory does not mean that an ethical theorist must be uncertain about the implications of their theory in practice. On the contrary, a theorist may sometimes say that the correct answer or response to a moral issue is obvious and certain. For example, they may insist that their moral theory definitely supports (or opposes) strong regulation of social media, protection of the property of artists when training Generative AI, the banning of autonomous smart weapons, or putting brakes on research programs to create human-level or super-intelligent AI.

But even when a clear answer is not forthcoming, the moral theorist will tend to say that their theory will in many cases give or point to an answer about what to do – and indeed, that giving clear ethical guidance about real-life questions and dilemmas is part of the very purpose of moral theory. Nonetheless, the high-level nature of moral theories does mean that some work will commonly need to be done to interpret and apply a theory to a specific, and often exceedingly complex and messy, real-world situation. Furthermore, thinkers who support a particular theory may well disagree on what 'answer' the theory will give to specific questions. Thus, virtue ethicists (for example) may disagree amongst themselves about, say, the extent to which we should regulate social media or research into human-level AI.

The generality of ethical theories is *'not a bug, but a feature'*. Theoretical frameworks for normative thinking are designed to capture basic criteria and fundamental standards or determinants of rightness and wrongness or of good and bad character. And they are meant to apply across all areas in which moral agents – individuals capable of moral understanding and moral action, that is, most adult humans – find themselves. These features are what gives ethical theories their enduring practical usefulness. Once we understand ethical theories, we can apply them to specific contexts. To use a popular current term, we can 'operationalise' the theory. In digital ethics, this means employing theories like, say, deontology or care ethics to address questions and situations involving digital technologies.

A final point about ethical theory should be made. Each theory effectively claims that it alone is the right theory and that all the other theories are wrong. Strictly speaking, competing moral theories are ultimately incompatible with each other. For example, you can't be both a committed utilitarian and a committed care theorist. That 'prohibition' might strike you as a touch arrogant! After all, didn't we talk a moment ago about the wisdom of approaching the difficult and often messy subject of ethics with some humility and open-mindedness?

However, we should appreciate that proponents of specific ethical theories have well thought-out reasons for supporting their preferred theory over others. Furthermore, moral theorists do not simply reject rival theories out of hand. On the contrary, they often try to accommodate insights from the other theories to strengthen their own. We can, then, learn something from each moral theory.

It is time to see what the four mainstream moral theories say. A great deal of ink has been spilt on each of these theories, so once again, we can only scratch the surface. Still, it will be possible to get the gist of them.

## Four Moral Theories

### Utilitarianism

Each ethical theory posits its own fundamental determinants or criteria of right action. What, then, are the basic criteria of right action for utilitarianism? It is actually rather simple; in fact, there is only one determinant or measure of rightness here. *The single criterion for utilitarianism is that an action (or omission) is right if and only if it produces the greatest possible utility.*[9]

Utilitarians call this master principle or rule the **principle of utility**. While on its face it is a simple rule, it needs some unpacking. There are two main elements of this theory to explain here. One is that utilitarianism tells us to produce the best possible consequences. The other is that these desirable consequences are to be explained in terms of 'utility'. That may all sound strange, but don't panic, it will soon make sense.

Let's start with the meaning of utility. Utilitarianism instructs us, as per the principle of utility, to maximise utility or produce the most utility that we can. But what is utility? Utility generally refers to what is useful; in utilitarianism, utility refers to what benefits us and makes our lives go well. In other words, utility for utilitarians is about individual wellbeing or happiness. For this reason, utilitarianism is often associated with English philosopher John Stuart Mill's *Greatest Happiness Principle*. According to Mill's principle, "... actions are right in proportion as they tend to promote happiness and wrong as they tend to produce the reverse of happiness".[10]

But now we can ask: What is happiness? For Mill – and his mentor and fellow Englishman Jeremy Bentham, who largely originated utilitarianism – happiness is, quite simply, composed of pleasure and the absence of pain. Thus, utilitarianism is about producing pleasures, such as physical and emotional enjoyment, and avoiding pains, such as physical suffering and emotional distress, in sentient beings who can feel those things. Some things might be conducive to pleasure, such as money. But money is not itself an experience and is therefore not intrinsically worthwhile. It is the pleasure, or the pain, that money brings that actually matters.

For Bentham, *pleasure is pleasure*, plain and simple. A given degree or intensity and duration of pleasure is equally good no matter what that pleasure derives from or consists in. Mill, by contrast, contended that some pleasures, even when they are of the same intensity and duration, have more utility or contribute more to happiness than other pleasures.

Mill believed that the pleasures of, say, intellectual activity are 'higher' and worth more than the pleasures of, say, eating and drinking. In contrast, on Bentham's way of thinking, 'lower' pleasures, like pleasure from playing simple children's games (or doomscrolling for hours on Instagram), are just as good as sophisticated pleasures like reading poetry or doing mathematics. Nonetheless, both Mill and Bentham held that pleasure is the measure of happiness or wellbeing, and hence of utility. This variety of utilitarianism is called **hedonistic utilitarianism**.

There are, however, other ways of measuring wellbeing or happiness. Although the contemporary Australian philosopher Peter Singer now agrees with Bentham, Singer is known for his earlier defence of ***preference utilitarianism***. According to this variety of utilitarianism, happiness or wellbeing is reducible not to pleasures per se, but to the satisfaction of preferences or desires. A person is thus better off not because they receive many pleasures and avoid pains, but because their preferences are satisfied (including their preferences regarding pleasure and pain).

For example, a marathon runner or a university IT student might prefer to endure significant pain or discomfort in the process of hard training or difficult learning, while receiving only a modest degree of pleasure. But because they strongly prefer these activities over more 'pleasurable' ones, satisfying preferences for marathon running or learning about IT contributes most significantly to the happiness of these individuals. The stronger the preference, the more that satisfying that preference contributes to happiness and the more that failing to satisfy it contributes to unhappiness.

Let's now turn to utilitarianism's interest in producing the most utility possible, that is, its interest in maximising utility (measured, as we said, in wellbeing or happiness of some sort). Utilitarianism determines right versus wrong action purely by whether the action – or omission, since not acting too can have results – produces the best possible consequences or the most utility. Crucially, only the *consequences* of the action or omission matter in determining rightness and wrongness for utilitarians. This means that the moral agent's intentions, motives, character, emotions, and feelings are all strictly irrelevant to the action's rightness. Utilitarianism is the most widely accepted variety of **consequentialism** – the theoretical view that rightness depends purely on the outcomes or consequences of what we do or don't do.[11] For consequentialism, including utilitarianism, the agent's mental state or character has no intrinsic bearing on rightness. For example, it does not matter for an action's rightness whether a person acts with kindness or with little feeling, as long as they maximise utility.

Some consequentialists believe that right action is that which produces some good consequences or a 'decent' level of utility or happiness. For example, a consequentialist might say that it is often 'good enough' to just make some people a *bit* less unhappy, even though one could do much more. Utilitarians don't believe that. Instead, utilitarians believe that the only actions which are right are those actions that *maximise* utility. The moral agent who fails to produce the most utility that they possibly can produce acts wrongly. You can see that utilitarianism sets a high bar for right action! Some critics think that having to always maximise utility, especially when it comes at great personal cost, is asking too much of human beings.

However, utilitarians point out that if it is right to produce some utility, then it is surely much better to produce more of it, and, moreover, that it seems wrong to not produce the most overall happiness when doing so is within our power. A world with more happiness, utilitarians remind us, is a better world, a better state-of-affairs, than one with less happiness and more unhappiness.

Because the principle of utility tells us to maximise wellbeing or happiness, we have to work out what such maximising looks like. For a start, greater happiness will result from more intense and longer-lasting pleasure (for hedonistic utilitarians) and from the satisfaction of stronger and more enduring preferences (for preference utilitarians). (*Compare the relative strength of a preference to eat a donut versus a preference to pursue a vocation as a computer scientist or a philosopher.*[12]) So we will need to take the *degree* or *strength* of pleasure, pain, or preferences within an individual into account when deciding what is right.

But in addition, we must also take into account *numbers*, that is, how many individuals might be affected. All things being equal, increasing the happiness of more rather than fewer individuals creates more utility and a better state-of-affairs, so we should do that. Indeed, Mill spoke of utilitarianism as aiming at the *'greatest happiness for the greatest number'*.[13]

Utilitarians also pay heed to the general *probability* of various action's leading to greater utility. Obviously, we will tend to produce more utility if we choose actions that are more rather than less likely to create happiness. Of course, often, it will not be certain or clear what outcomes will result from various behaviours. In many cases, we just have to use our best judgement, to try to bring about the most happiness or wellbeing.

Utilitarians must also weigh and balance pleasures and pains or preferences. Pleasure counts for positive utility, pain for negative utility (or 'disutility'). Similarly, preference satisfaction promotes utility, while preference dissatisfaction diminishes utility. To produce the best consequences, we need to subtract negative utility (disutility) from positive utility to calculate the *sum-total of utility*. The sum-total of utility may include a wide array of possible pleasures, pains, or preferences, and these may have different strengths within and across individuals.

Consider, for example, the policy of banning foreign-owned popular social media platforms from harvesting personal data from users in other countries or jurisdictions. Banning the platform has complex implications, including various possible benefits for some and harms for others. Again, we need to use our best judgement in performing the *utility calculus*, since things are often complicated in socio-technical systems and life generally.

Finally, utilitarianism counts all similar pleasures, pains, and preferences – or *interests* as utilitarians sometimes call them – as equal in the utility calculus. A pain of a certain intensity and duration has the same value no matter who's pain it is. That includes the moral agent themselves, and all other interest-bearers, no matter if they are Black or White, male or female, old or young, poor or rich, and so on. The similar interests of all human beings count equally.

Indeed, so too do the interests of sentient nonhuman *animals*. Many animals also can experience pleasure and pain and have preferences. As a matter of fact, the utilitarian rule of equality would apply to any alien being, or future AI or robot, that have interests (if they do). Peter Singer talks about the equal consideration of similar interests for sentient beings, human and nonhuman. Singer's rule brings out an important part of utilitarianism's principle of utility.

## Deontology

In contrast with utilitarianism and its master principle of utility, deontology has a very different criterion of right and wrong action.[14] According to deontology, the right action(s) is that which accords with certain basic obligations or duties (*deontic* means duty). Wrong actions fail to accord with those obligations. The focus thus switches from a sole pre-occupation with consequences to a close attention to duties.

In one respect, deontology is more complicated than utilitarianism. While utilitarianism recognises only one ultimate imperative – namely *maximise utility* – forms of deontology recognise several core duties. These duties cannot be reduced to other more basic duties. Instead, these duties *are* the most basic duties – they give the ultimate reason for judging actions right or wrong.

But what are these basic duties? That depends on the type of deontological theory. The British philosopher W.D. Ross is known for his theory of **prima facie duties**. We reproduce a list of his duties here:

1. *Fidelity.* We should strive to keep promises and be honest and truthful.
2. *Reparation.* We should make amends when we have wronged someone else.
3. *Gratitude.* We should be grateful to others when they perform actions that benefit us and we should try to return the favour.
4. *Non-injury (or nonmaleficence).* We should refrain from harming others either physically or psychologically.
5. *Beneficence.* We should be kind to others and try to improve their health, wisdom, security, happiness, and well-being.
6. *Self-improvement.* We should strive to improve our own health, wisdom, security, happiness, and well-being.
7. *Justice.* We should try to be fair and try to distribute benefits and burdens equably and evenly.[15]

Do these strike you as basic or ultimate duties? For most of us, this list will probably seem plausible and perhaps even self-evident. We might perhaps question whether we have a moral duty to ourselves to make efforts at self-improvement, as opposed to, say, thinking it merely prudent and wise. For example, we might think we cannot wrong ourselves by spending all our time video-gaming instead of also undertaking other activities, even though others might think it an unfortunate waste of our talents. Still, many or all the duties on Ross's list might seem very reasonable to most of us, even when we come from different societies and cultures.

One of the complications of deontology is that its core duties can conflict. Consider an automated resumé-sorting system. Some companies use these systems so that people can avoid the time-consuming and boring job of scanning hundreds or thousands of resumés, a task that diverts them from more interesting and productive tasks for the company. Using an AI program for resumé screening apparently accords with the duty of *beneficence*, since workers and company stakeholders may benefit from it.

However, as seen in Chapter 3, we know that even when great care is taken in training AI, such machine learning programs can remain biased against certain resumés, such as those from women, carers, minorities, and others. Employing the tools for convenience and efficiency thus seems quite unfair. That is, using this AI resumé-sorting system may accord with the duty of beneficence but *clash* with the duty of justice.

What do we do when conflict between duties occurs? In Ross's view, these duties are not absolute, not to be followed no matter what; they are rather, as he puts it, prima facie duties. *Prima facie* is Latin for 'on the face of it'. Prima facie duties give us basic guidance and reasons for actions, but they can be overridden by other prima facie duties. Navigating conflict in duties requires our judgement and reasoning in specific cases.

In the resumé-sorting example, we may judge that justice to women seeking work is a more pressing or a stronger duty than beneficence to company workers and stakeholders. Women and other groups have also been historically discriminated against in certain workplaces and society. As such, we might conclude that the duty of justice should override the duty of beneficence in this instance. But that will not always be so, and sometimes beneficence (or some other duty) will override justice. It depends on the circumstances, our judgement in applying the basic duties, and the supporting reasons we give.

We said that according to deontology, rightness depends on acting in accord with duties that give us reasons for action. To act rightly involves more than just doing the things that the duties prescribe. It can also mean acting with a certain *intention and motivation*. For example, acting justly or fairly in the resumé-sorting case does not merely involve rejecting the biased AI system because, for example, it is too expensive to subscribe to. Rather, it means rejecting the system because it is unfair and fails the *duty of justice*. In this further way, deontology is different from utilitarianism, since that theory says a moral agent acts rightly as long as they maximise utility, even if that is not their intention.

The idea that acting rightly depends on our reasons for action and obedience to duty also comes out (very forcefully) in a different form of deontology, that of German philosopher Immanuel Kant. Kant's highly influential approach to ethics is strongly opposed to utilitarianism. Kant argued that rightness requires acting on what he called the *moral law* (not to be confused with human or state laws) rather than promoting utility. Because his moral philosophy is highly complex, we will simply give you a broad overview of some essentials.

Morality, it would appear, does not make exceptions for individual moral agents. For example, as a moral agent I cannot personally assert that every other moral agent should be kind to me, while simultaneously asserting that I can be unkind to others whenever the mood takes me. Kant claimed that the demands or commands of morality – or moral 'imperatives' as he called them – apply equally to each and every moral agent. They are universal. And these imperatives apply to moral agents unconditionally – or 'categorically' – regardless of an agent's own, often highly changeable, interests, goals, and feelings.

Kant argued that we moral agents can recognise that morality issues a **categorical imperative**. A categorical imperative is an unconditional command that we must follow on pain of irrationality and irrespective of our individual feelings and goals. There are two key formulations of the categorical imperative. The first is the *Formula of the Universal Law of Nature*. It goes like this:

> Act only in accordance with that maxim [or rule or principle] through which you can at the same time will that it become a universal law.[16]

This is tricky, so let's spell things out. The moral command in that sentence can be thought of as a rational test of whether an action is wrong or permissible. If we want to act on a certain maxim or rule, we can check to see if that rule passes or fails the test of the categorical imperative. This means we need to *universalise* the rule: We need to apply it to all moral agents.

Suppose I am considering acting on the following rule of action: *'Lie to other moral agents when it suits my interests'*. To determine if this behaviour is morally wrong or acceptable, I need to see if I can rationally implement this rule as a universal law. That means I must think about what would happen if other moral agents were to accept and act on that rule too.

What would happen if everyone acted on my universalised rule 'lie to other moral agents when it suits my interests'? If we think about this, we can see that people would begin not to believe one another. As lying becomes rampant, no-one would believe me or you or anyone else about anything. And if that calamity happens, and the whole idea of truth-telling breaks down, it would be impossible for me to lie, since lying depends on other agents believing my deceptions.

The same problem afflicts the action of making false promises. If false promising became a universal practice, no-one would accept promises, so it would be impossible to make them, even falsely. Something similar applies to stealing. A world in which everyone steals when they want to would be a world in which the institution of property ownership breaks down.

I cannot rationally implement my rule about lying (or false promising, or stealing) as a universal law or universalised rule. In fact, when I aim to lie, I effectively defeat my purposes and contradict myself. Lying fails the rational test laid down by the first formulation of the Kantian categorical imperative. Kant concludes that lying and false promising – both forms of deceiving others – are therefore always morally wrong. Importantly, lying is not wrong on Kant's view because it would harm society or reduce happiness, as a utilitarian might say. Rather, lying is morally wrong because in seeking to lie I contradict myself and act irrationally.

Kant held that applying this test would tell us more generally which actions are wrong and which permissible. The categorical imperative has various implications for digital ethics. For example, the categorical imperative could explain why it is wrong for platforms or apps to deceive users about collecting personal data and selling it to third parties, or to deceive them about the targeted nature of the ads they receive based on user profiling.

Kant also produced a second formulation of the categorical imperative which he claimed was another version of the first. Moral agents, said Kant, are capable of understanding and acting on the moral law, on the categorical imperative encountered above. We just saw that the moral law entails the wrongness of deceiving other moral agents. Indeed, it would be always and absolutely wrong, without exception, to set out to deceive others.

Lying and making deceitful promises to other agents are ways of manipulating and using them for our own goals. For Kant, such use of other moral agents is treating them as mere means to our personal ends and is absolutely wrong. This leads us to the second formulation of the categorical imperative, or *The Humanity Formula*:

> So act that you use humanity, whether in your own person or in the person of any other, always at the same time as an end, never merely as a means.[17]

By 'humanity', Kant meant moral agents – beings that can rationally grasp and act on the moral law. Beings with that ability have *autonomy*. They do not just act on various feelings, beliefs, and goals but can act on considered moral reasons. Now such beings, in Kant's view, must always be treated as *ends in themselves, never as mere means* to other ends. It is thus always wrong to use them by riding rough-shod over their autonomy. Kant claimed that autonomous beings have a *dignity*. He maintained that it is absolutely wrong to kill innocent agents who possess autonomy (though he favoured capital punishment of the guilty).

An interesting feature of Kant's view is that moral agents need not be human. Therefore, if there are, now or in the future, other animal species, aliens, AIs, or intelligent robots that are autonomous moral agents, these entities must also be treated as ends in themselves, and on no occasion used as mere means.

## Virtue Ethics

Both deontology and utilitarianism lay down explicit duties or principles (or a single principle in the case of utilitarianism) for moral agents to follow and carefully apply in various real-life situations. Virtue ethicists claim things are not so straightforward.[18] Morality, say virtue ethicists, is more complex. Rather than giving us explicit principles and duties, virtue ethics has it that the criterion for right and wrong action is this: *Do what a truly virtuous person would do (and don't do what a virtuous person would not do)*. This will need some unpacking.

We can start by asking what a truly virtuous person is. A truly virtuous person is a morally good person who possesses the virtues. What are virtues? The **virtues**, according to this moral theory, are settled ethical traits or dispositions, tendencies embedded in our characters to act and to feel in certain ways. Examples of virtues include courage, justice, temperance (or self-control), honesty, compassion, and generosity. A truly virtuous person possesses these virtues to an impressive and exemplary degree. Such a person is eminently worthy of our moral respect. In fact, we should ourselves seek to be like the person of real virtue.

The opposite of a virtue is a **vice**. Vices correspond to contrary virtues. Thus, some corresponding vices of the virtues we just listed are cowardice, injustice, intemperance (lack of self-control), dishonesty, callousness, and stinginess. A non-virtuous or vicious person has a character that leads them to practice vice rather than virtue, and so to routinely act wrongly. Such a character is morally bad and corrupt, and certainly not worth emulating. Remember when we mentioned the *twin vices of moral complacency and panic* in our Preface? Avoiding those vices would mean being morally proactive yet collected and thoughtful.

According to virtue ethics, virtues are highly context specific. Aristotle, the famous ancient Greek philosopher who originated virtue ethics, put it like this: A virtue involves acting and feeling in the right way, at the right time and place, to the right degree, and towards the right objects.[19] Let's take the example of honesty as a virtue. The truly honest person will practice honesty not in any old manner or willy nilly but in a certain way and in the right circumstances.

An honest person will, of course, tend to tell the truth. They will also be pained when they encounter lying. But being virtuous can get complex. Suppose our honest and virtuous person works for a tech company. This honest person will not blurt out company secrets to competitors who ask for them. We might think that someone who was honest must tell the truth to anyone who asks for it. Yet doing that would not show true or genuine honesty. It would rather indicate a distorted view of truth-telling.

On many other occasions, however, the virtuous person *would* tell the truth, as long as the conditions are right and require truth-telling. Such a requirement could even occur when telling the truth could damage the company's reputation. For example, if the tech company is causing serious social harm, the honest person might break company secrecy and inform the public and lawmakers to prevent ongoing damage to people.

This is what happened with Facebook. Several years ago, Facebook whistleblower Frances Haugen, as we have mentioned in Chapter 1, told US politicians about the company's use of an algorithm that maximised user engagement but harmed teens.[20] In going public, Haugen also arguably displayed other virtues. For example, standing up to a tech titan involves a great deal of courage. Haugen was also motivated to speak out due to a sense of justice and compassion for the victims of Facebook's greedy and callous business model.

Now virtues involve not only right action but *right feeling* – feeling the right things in the right circumstances. This feature further differentiates virtue ethics from utilitarianism and deontology. The honest and just person is pained by dishonesty and injustice. Or to take another example, the virtue of courage involves feeling the right amount of fear and confidence in the right situations. For example, when it would be crazy and reckless to walk into a highly dangerous situation for no good reason, it would be quite right to feel fear about such a course of action.

The courageous person, however, will sometimes master their fear when the circumstances require it, and perform the dangerous activity. Overcoming fear and acting courageously could, as we just saw, involve publicly blowing the whistle when a tech company is acting egregiously. Or it could involve something more ordinary, like standing up for an unfairly treated work colleague.

Some people wonder whether the virtues really are of much help as the criterion for right action. For how can virtues guide our behaviour when we need to already be virtuous – to be an exemplary moral agent – in order to know when to perform certain actions and have certain feelings? If we are not already genuinely and highly virtuous, how can we identify virtuous people? Aristotle had some relevant things to say about this worry.

Firstly, Aristotle acknowledged that there is some luck involved in being able to identify and practice virtue. We must have been brought up with some good guidance from our parents and community to avoid becoming vicious and to appreciate virtue in others. Indeed, Aristotle claimed that attaining virtue is partly a matter of good training. A person who lacks training and who has already become vicious and bad, has little hope of identifying and practicing virtue. But that is (hopefully!) not the case for most of us.

Furthermore, by *regularly trying* to behave virtuously, and by befriending and being around virtuous people, Aristotle claimed that we can acquire the habits of virtue. If we want to become virtuous digital practitioners, then we should choose our company – perhaps in both the friend-related and the corporate senses of '*company*'! – wisely. If the company we keep is callous and cruel, or dishonest and ungenerous, we may gradually become like them. We should also practice being courageous, honest, self-controlled, just, compassionate, and so on. With practice and repetition, being virtuous, Aristotle thought, becomes more effortless and even pleasurable.

Secondly, the virtues for Aristotle have a certain structure (Table 4.1). Virtues can go awry in two ways: through *excess* and *deficiency*. That is, vice can result from either an excess of certain action and feeling, or from a deficiency in them. Take again the virtue of courage. An deficiency of action can mean not being doing what is required in the right circumstances and thus showing the vice of *cowardice*. An excess of action, however, can also undermine courage and cause a different vice: *recklessness*.

**TABLE 4.1**

Examples of Some Virtues and Vices[21]

| Virtue | Vice: from deficiency | Vice: from excess |
|---|---|---|
| Courage | Cowardice | Recklessness |
| Generosity | Stinginess | Wastefulness |
| Compassion | Callousness | Condescending pity |
| Honesty | Deceitfulness | Brutal candour |
| Temperance | Overindulgence | Insensitive asceticism |
| Right temper | Apathy or insufficient anger | Hot-tempered |
| Benevolence | Malevolence | Overbearing altruism |

For Aristotle, identifying and exhibiting virtue requires avoiding both excess and deficiency in action and feeling. What is virtuous can vary greatly with circumstances. Thus, in some circumstances, it is right and appropriate to have very strong feelings, and in other cases, it is not.

For example, if an automated AI decision-making system that allocates welfare payments were to deny many unemployed people the money essential for daily living, it might be virtuous to feel moral outrage towards the authorities running such a system. Yet if a flawed welfare allocation system were to only make small and rare errors, it might be appropriate to feel a more modest amount of anger or no anger. Aristotle says that virtues involve aiming at and hitting a mark – sometimes called the *'golden mean'* – between deficiency and excess. This mark or golden mean will vary with the specific circumstances the moral agent is in.

Since virtue is so contextual, knowing how to hit the target or mean requires gaining a certain skill. Aristotle referred to the skill, indeed the excellence, of *practical wisdom*. Practical wisdom (or *phronesis* in Aristotle's Greek word) is the skill of knowing how to hit the 'mean' between excess and deficiency. Only then can we successfully achieve virtue. But doing that is not easy, or even natural. Aristotle claimed that one only gains excellence in this ability through extensive training and habit, and by learning from impressively virtuous people.

The philosopher Shannon Vallor has applied virtue ethics – drawing on both Aristotle and non-Western philosophies[22] like Buddhism and Confucianism – to digital ethics. Vallor believes that our lives are now so entangled with forms of digital technology that our virtues effectively need to be *techno-moral* virtues. She claims we need techno-moral virtues to navigate these fascinating but fraught times of rapid technological progress.

Indeed, we now live in a world where technology increasingly makes sensitive decisions about our lives and wellbeing. Furthermore, some technologists, rightly or wrongly, will try to make technology that could do quite radical things, like profoundly enhancing our abilities, intelligence, and life spans. Perhaps most significantly, digital technology could either exacerbate or help mitigate disastrous climate change. Vallor argues that we – everyone who uses technology and the people who create and operate it – need to develop *techno-moral wisdom*.[23]

This term echoes Aristotle's idea of practical wisdom. But Vallor's 'techno-moral' wisdom is informed by the special nature and implications of digital technologies, which Aristotle could never have imagined. For instance, in our current situation – where things may soon get very much worse for the planet and its human and animal inhabitants – we desperately need virtues like courage, justice, compassion, and hope. But we also need the skill and wisdom to recognise when and how various kinds of technology could improve or worsen our lives and our present predicament.

## Care Ethics

The fourth and final ethical theory that we will discuss is care ethics, sometimes also called the 'ethics of care'.[24] Care ethics developed as a significant philosophical approach beginning from the 1970s – far more recently than the previous theories. With care ethics, we fortunately begin to see a greater role being played by women thinkers because of welcome cultural changes. This is in contrast with the historical development of utilitarianism and deontology.

To be sure, virtue ethics has attracted a number of significant female philosophers, including Philippa Foot, Elizabeth Anscombe, Rosalind Hursthouse, and, in the digital space, Shannon Vallor. Yet virtue ethics' founding father was Aristotle. In contrast, it was female thinkers who were responsible for the initiation and original construction of an alternative 'care-based' approach to ethics. Indeed, care ethics takes off from a position of feminist criticism of earlier theories. Key feminist thinkers in care ethics' development include Nel Noddings, Sara Ruddick, Virginia Held, Eva Feder Kittay, and Joan Tronto.

These philosophers contended that mainstream theories like utilitarianism and deontology tended to issue rules, commands, or principles that are (a) highly abstract and removed from human relationships and communities and (b) largely divorced from ways of feeling and understanding through emotional responses and attachments. These feminist critics suggested that such theories represent stereotypically masculine ways of thinking. Dominant male perspectives, they argued, shoulder out alternative, legitimate viewpoints, which are more often held by women (though can equally be held by men).

A seminal moment in the development of care ethics involved the late 20th century work of moral psychologist Carol Gilligan, captured in her book *In a Different Voice*.[25] Gilligan observed that boys and girls may respond differently to moral questions or dilemmas. In a famous study, children were asked to comment on whether it would be right or wrong for a man to steal from a pharmacist a drug which he could not afford, to save his dying wife. Boys tended to justify the theft in terms of concepts such as justice, rights, and wellbeing. In contrast, some girls pointed to how the theft might affect the relationship between husband and wife (e.g., if the man ends up in jail) and suggested ways of negotiating relations with the pharmacist, perhaps explaining to them the situation of the dying wife.

Boys thus more often opted for the impartial, abstract principles found in established moral theories, whereas some girls tended to recognise ethical notions involving relationships, empathy, and care. Although male psychologists concluded that such girls were therefore less morally advanced than boys, Gilligan contended that 'non-masculine' approaches were *different* rather than inferior.

Care ethicists have since argued that other moral theories ignore what is in reality most basic and central in moral life. Although care ethics is not monolithic, its adherents point to some vital features of morality that this approach captures. Firstly, we do not normally relate to all other human beings, as utilitarianism and deontology tend to suggest, as strangers to whom we owe impartial duties of, say, justice or utility promotion. Rather, we are involved in a variety of caring relationships: mother and baby, parent and child, brother and sister, friendships, neighbours, fellow members of religious or other groups, and so on. Some of our core responsibilities involve forming and maintaining these relationships and bonds. We thus have special obligations to certain individuals over total strangers, and the nature of our obligations are different than may be supposed by more 'abstract' approaches.

Secondly, *these relationships are not affectless or cold*. Relationships and the care internal to them are constituted by felt responses such as attachment, empathy, love, and compassion. A person suffering from psychopathy would be incapable of understanding the ethical nature of these relationships, even if they could follow certain rules. Lacking these caring emotions, such a person would also necessarily be unable to uphold obligations of care toward others. Care ethicists may certainly acknowledge that emotions are insufficient on their own; for example, caring responsibly requires practical skills, such as being able to competently identify precisely how to meet others' needs. Nevertheless, many care ethicists would contend that properly attending and responding to others requires some ability to feel.

Thirdly, human beings are *vulnerable, embodied beings*. On traditional ethical theories, perhaps most conspicuously Kant's deontology, human moral agents are depicted as fully formed autonomous beings. This, say care ethicists, is an illusion. All human moral agents were once helpless babies, and all these agents will eventually die (unless, by some miracle, AI finds a way of making humans immortal and immune to accidents!). Furthermore, since we are biologically embodied creatures, even the most self-sufficient human moral agent could at any moment suffer an injury or disease that renders them non-autonomous and entirely dependent on the care of others.

Thus, care ethicists argue, we should relate to and care for one another not as invulnerable and forever-autonomous agents, but as fragile and embodied human animals. Of course, some members of society have been relatively privileged and lucky. Others have been less so. For example, some people have physical and intellectual disabilities and face associated discrimination. Others, such as ethnic minorities, Indigenous people, women, and members of LGBTIQA+ communities, have suffered historical oppression and endure ongoing social prejudice. Some individuals have much less power than more fortunate members of society.

As with other ethical theories covered in this chapter, care ethics encounters some criticisms. One possible criticism is that the approach could reinforce essentialist gender stereotypes; another is that care ethics cannot provide clear enough guidance about how to act, perhaps especially towards individuals to whom we are not attached.[26] Nonetheless, care ethics could be a useful approach both in its own right, and, perhaps, as a means of providing an alternative perspective to the rationalistic, 'scientific' perspectives ascribed to technology at large – perspectives that might emphasise, say, unfeeling calculation over our entanglements with and human responses towards others.

The above features of care ethics have potentially important implications for digital ethics. For example, care ethicists will claim that decisions to create and use powerful technologies should consider how they may either damage or nurture human relationships and associated responsibilities of care. Decisions to make impactful technology, these ethicists may say, should also pay particular attention to the more vulnerable and needy members of the community, rather than, as is often the case, catering to the less vital needs of more powerful and lucky members – including those who are less vulnerable to the negative effects of technology and who have the means to access beneficial technologies. Even though the latter too may become needy and dependent, it is the former that most urgently need society's help and protection.

* * *

At this point, readers might have some questions. *How do these theories apply in practice, including to uses of digital tech? What happens when one theory seems to tell us to do one thing, while another seems to tell us to do a contrary thing*? You might also wonder why we have only touched on four main ethical theories from the Western philosophical tradition or *canon*. *Aren't there more ethical approaches in the world*?

With these questions in mind, the next chapter will apply what we have learnt thus far to a practical problem. We shall also see if there are other approaches which might help us make sense of ethical conundrums, beyond the four moral theories we have encountered.

## Notes

1. Pojman (1995).
2. To recap, see The Facebook Files (2021, October 1).
3. Australia is one of these jurisdictions. For further reading, see Australian Human Rights Commission (2024, November 20).
4. In popular culture, the Trolley Problem (deriving from philosopher Phillipa Foot) is just one example of such dilemmas and forms the foundation for various plot devices. Examples include the movie *Sophie's Choice* (1982), the TV series *The Good Place*, as well as games such as "Would You Rather …".
5. Awad et al. (2018).
6. Wolf, Miller & Grodzinsky (2017).
7. Lee (2016, March 25).
8. Cadwalladr & Graham-Harrison (2018, March 17).
9. de Lazari-Radek & Singer (2017).

10. Mill (2019).
11. Sinnott-Armstrong (2023).
12. Interestingly, more thought-experiments involving comparisons such as these – but on a broader scale, to decide if *'ethics* itself is even worth it' – can be found in Chapter 6.
13. Mill (2019).
14. Wood (2008).
15. Simpson (n.d.).
16. Kant & Ellington (1993).
17. Kant & Ellington (1993).
18. Swanton (2003).
19. Aristotle & Ross (2009).
20. For further reading, see Haugen's (2023) memoir, as briefly mentioned in Chapter 1.
21. Swanton (2003), Vallor (2018).
22. Non-Western philosophies will be expanded upon in Chapter 5.
23. Vallor (2018).
24. Held (2006).
25. Gilligan (1993).
26. Sander-Staudt (n.d.).

## References

Aristotle, & Ross, W. D. (2009). The Nicomachean Ethics (W. D. Ross & L. Brown, Trans.). Oxford University Press.

Australian Human Rights Commission. (2024, November 20). Proposed Social Media Ban for Under-16s in Australia. https://humanrights.gov.au/about/news/proposed-social-media-ban-under-16s-australia

Awad, E., Dsouza, S., Kim, R., Schulz, J., Henrich, J., Shariff, A., Bonnefon, J.-F., & Rahwan, I. (2018). The moral machine experiment. Nature, 563(7729), 59–64. https://doi.org/10.1038/s41586-018-0637-6

Cadwalladr, C., & Graham-Harrison, E. (2018, March 17). Revealed: 50 Million Facebook Profiles Harvested for Cambridge Analytica in Major Data Breach. The Guardian. https://www.theguardian.com/news/2018/mar/17/cambridge-analytica-facebook-influence-us-election

de Lazari-Radek, K., & Singer, P. (2017). Utilitarianism: A Very Short Introduction. Oxford University Press.

Gilligan, C. (1993). In a Different Voice: Psychological Theory and Women's Development. Harvard University Press.

Haugen, F. (2023). The Power of One: How I Found the Strength to Tell the Truth and Why I Blew the Whistle on Facebook (1st ed.). Little, Brown and Company.

Held. (2006). The Ethics of Care: Personal, Political, and Global. Oxford University Press.

Kant, I., & Ellington, J. W. (1993). Grounding for the Metaphysics of Morals; With, On a Supposed Right to Lie Because of Philanthropic Concerns (3rd ed.). Hackett Pub. Co.

Lee, P. (2016, March 25). Learning from Tay's Introduction. The Official Microsoft Blog. https://blogs.microsoft.com/blog/2016/03/25/learning-tays-introduction/

Mill, J. S. (2019). The Greatest Happiness Principle: Utilitarianism, on Liberty & the Subjection of Women. Madison & Adams Press.

Pojman, L. P. (1995). Ethical Theory: Classical and Contemporary Readings. Wadsworth Publishing.

Sander-Staudt, M. (n.d.). Care Ethics. Internet Encyclopedia of Philosophy. Retrieved March 25, 2025, from https://iep.utm.edu/care-ethics/

Simpson, D. L. (n.d.). William David Ross (1877–1971) [Internet Encyclopedia of Philosophy]. Internet Encyclopedia of Philosophy. Retrieved March 25, 2025, from https://iep.utm.edu/ross-wd/

Sinnott-Armstrong, W. (2023). Consequentialism. In E. N. Zalta, & U. Nodelman (Eds.), The Stanford Encyclopedia of Philosophy. Metaphysics Research Lab, Stanford University. https://plato.stanford.edu/archives/win2023/entries/consequentialism/. https://plato.stanford.edu/archives/win2023/entries/consequentialism/

Swanton, C. (2003). Virtue Ethics: A Pluralistic View. Oxford University Press.

The Facebook Files. (2021, October 1). Wall Street Journal. https://www.wsj.com/articles/the-facebook-files-11631713039

Vallor, S. (2018). Technology and the Virtues: A Philosophical Guide to a Future Worth Wanting (First Issued as an Oxford University Press Paperback). Oxford University Press.

Wolf, M. J., Miller, K., & Grodzinsky, F. S. (2017). Why we should have seen that coming: Comments on Microsoft's Tay "experiment," and wider implications. SIGCAS Computers and Society, 47(3), 54–64. https://doi.org/10.1145/3144592.3144598

Wood, A. W. (2008). Kantian Ethics. Cambridge University Press.

# 5

# *Being (Digitally) Ethical*

## Introduction

In an important sense, ethics is 'not a spectator sport'. As we saw, Aristotle believed that practice is essential to our ethical development as individuals. Aristotle thought that the more we strive to be ethical, the more we acquire moral habits and skill (so long as we have the right guidance). In addition to trying to become better people in our daily lives, we can also practice using ethical concepts and frameworks on case studies and examples.

To illustrate, let us exercise our ethical thinking muscles with a rather plausible ethical controversy – one that arises from present technological capabilities and that may well become more common, to greater or lesser degrees. The case study we will explore is the use of artificial intelligence (AI) in the workplace to monitor worker productivity. We will examine what each of the previously discussed moral theories – utilitarianism, deontology, virtue ethics, and care ethics – may have to say about it.

## Moral Theories in Action: Workplace Monitoring

You might have heard that some digital technologies (with or without AI) have been deployed by employers to track employees' productivity, especially during work-from-home periods during the COVID-19 pandemic.[1]

Now let's imagine the following hypothetical scenario:

> A food company, let's call it *FastFresh*, has a distribution centre for its goods. The focal point of this centre is the warehouse, where workers pick requested items from shelves and load them onto trolleys, to be distributed to online shoppers. Video cameras surveil the warehouse, and an AI system measures worker speed and accuracy and determines a productivity rating for each worker. Management of FastFresh can use these ratings to warn, counsel, and even retrench workers who are significantly slower or less accurate than average.

DOI: 10.1201/9781003393511-5

Before ethically exploring this scenario, some preliminary points should be made to avoid misunderstanding. Firstly, a given moral theory may be applied to a situation in *different* ways. Different thinkers, even when they are experts in that moral theory, may be at odds over what moral conclusions should be drawn from a moral theory in action. This can be due to differing judgements about the *facts*, about the particular *theory* or various versions of it, and about *how* to apply the theory.

Secondly, while rival ethical theorists may sometimes arrive at different moral conclusions, they may often reach the same conclusions about what to do. However, even when rival theorists broadly agree about should be done, they may do so for very different reasons. Moreover, they may disagree about what rightly doing a given act comes to. For example, a utilitarian would say that as long as utility is maximised, then the right action has been done, regardless of the motivations or the reasons the moral agent has for doing it. But other theories strongly disagree with that, and each in their own distinctive way.

Thirdly, our application of ethical theories is abbreviated, simplified, and selective. There is much more that could be said in each case, but we are limited by space constraints and the complex nature of such moral exercises. With these provisos out of the way, let's return to the case study. Our question is, therefore:

> Is use of the AI system by the management of FastFresh ethically acceptable?

## The Utilitarian Says ...

Consider first how utilitarianism might deal with the ethics of this case of workplace monitoring. Utilitarianism tells us to sum up positive and negative utility, or intrinsically valuable states of affairs, in the form of experiences or preferences that result from available actions.[2] There are clearly important sources of negative utility here in the form pain and preference dissatisfaction. Employees at FastFresh will likely feel very uncomfortable being monitored and rated against each other. This may significantly impact on their mental wellbeing. Rushing to lift and place heavy items may also cause workplace injuries. Deployment of monitoring systems may even prevent employees going to the toilet and taking breaks, for fear of receiving lower productivity ratings. Furthermore, workers may suffer substantially and for extended periods if they are reprimanded and fired for being slower and less accurate than fellow workers.

However, there are also sources of positive utility. Assuming the AI system works, greater productivity boosts company profit, benefitting shareholders by increasing their pleasure or satisfying their desires. Shopping customers also experience more pleasure and preference satisfaction from rapid food delivery with fewer mistakes in their orders.

The principle of utility stipulates that the right action maximises utility. Would utility be maximised by using, or rather by not using, the AI system at FastFresh? Now, as we saw, using the system generates negative utility for employees, which counts against it. However, the system also generates positive utility for shareholders, which counts for it. We must now try to determine whether the sum-total of utility for deploying the technology is greater than shelving it. To do this (at least in simplified way), we will need to make some assumptions. It seems reasonable to assume that the resulting health and stress impacts on employees is significant. The intensity of the pain and the strength of desires of employees not to be subject to the AI system means it generates quite a lot of disutility. If the AI system was not deployed, this disutility could be avoided.

Plausibly, the significant pain suffered by an employee outweighs the moderate pleasure gained by a company shareholder from the somewhat greater profits brought about by efficiency gains. The pain also surely outweighs the increased convenience for an online food shopper. However, the principle of utility also takes *numbers* into account. Suppose for argument's sake the company employs 1000 workers and that there are 500 shareholders and 100,000 online shoppers. Although the impact of the AI system generates significantly more disutility for a single employee than it generates utility for a single shopper, there are many more shoppers than employees.

So, though it is not certain, the total of this utility *might* exceed the total disutility of the AI use. This means that it is at least possible that the option of installing the AI system brings about the greatest sum-total of utility than other available options. If that is true, then utilitarianism may support the use of the AI system. If not, then it would oppose it.

In reality, a utilitarian may find that there are sound consequentialist reasons to oppose this AI system. For example, it may be that the promised efficiency gains can be brought about not with the stick of AI monitoring but with the carrot of improved pay and conditions for workers. In that case, greater utility would result from a course of action that avoids using this AI surveillance system. Utilitarians may also be rightly concerned about setting precedents that lead to other forms of workplace or societal monitoring that decrease people's happiness. However, it is important to note that the utilitarian should at least be *open* to finding that this type of surveillance and rating of workers is ethically acceptable, and even obligatory *if* it maximises overall happiness compared to alternative actions.

### The Deontologist Says ...

In applying deontology to our case study, let's start with W.D. Ross's prima facie duty[3] account (Chapter 4). We might first identify which prima facie duties are relevant here. It seems that duties of nonmaleficence (or avoiding causing harm), beneficence (or doing good), and justice all fit our case.

For example, there is a prima facie duty to not cause pain and distress to the employees of FastFresh. However, managers of the company also have a prima facie duty to benefit company shareholders. Recall that a prima facie duty functions as a reason for action but *can be overridden* in circumstances where there is a more pressing or stronger prima facie duty that conflicts with it.

As we just saw in discussing a possible utilitarian response, there are many more customers who would benefit than there are employees who would be harmed in this scenario. Perhaps, then, the duty of beneficence (to customers) should override the duty of nonmaleficence (to workers) in this situation, making the AI system morally acceptable. However, the setback in the interests suffered by workers (who could face severe stress and job loss) is surely much weightier than the benefits that accrue to shareholders and shoppers. Thus, a deontologist might decide that FastFresh management's duty of nonmaleficence towards the employees overrides their duty of beneficence to shareholders and customers.

Notice that our deontologist is not bound to agree with the utilitarian, since they do not subscribe to the principle of utility. Accordingly, the deontologist may reasonably claim that the duty of nonmaleficence here is stronger than the duty of beneficence, even if acting in the former way does not maximise utility. In fact, there is another way that the deontologist may be at odds with the utilitarian: The deontologist may judge that the prima facie duty of *justice* not only counts against the FastFresh AI system but is a decisive moral objection to it.

Let's explore this. Using the AI system, the deontologist may contend, makes the company and its shareholders somewhat richer, and improves the online shopping experience. But it does so on the backs of relatively low-paid employees whose lives are made relatively miserable. The system is therefore quite unfair to employees. It is worth noting that a key objection to utilitarianism, from deontologists as well as from other theorists, is that the theory countenances some acts of injustice. For their part, utilitarians typically reply that the non-utilitarian conception of justice gives insufficient weight to increasing happiness and reducing suffering.

The objection from justice in our scenario arguably gains additional force from the Kantian version of deontology. Recall the second formulation of Immanuel Kant's categorical imperative. This commands us never to treat another autonomous moral agent as a mere means but rather only and always as ends in themselves.[4] We may never, the categorical imperative implies, treat moral agents as mere means even if the negative consequences of not doing so are unfortunate or even dire. Arguably, overriding the autonomous wishes of FastFresh employees to not be subject to constant technological monitoring and evaluation is an example of treating them as mere means for the company's ends. We can easily imagine employees protesting to management: *"You are just using us as your tools! We deserve to be treated with dignity and justice!"*

## The Virtue Ethicist Says ...

The next theory to apply to our case study is virtue ethics. It seems possible that a virtue ethicist might also make a case that FastFresh's AI system is morally unacceptable. Remember that virtue ethics' criterion of rightness is what a person of exemplary virtue would do and feel in the relevant circumstances.[5] How would such a person react to the case study? The virtuous person, to be sure, need not ignore the benefits to shareholders and customers. In some cases, it may show the virtue of benevolence to seek to benefit these parties. But there are other virtues to consider in this situation.

Prime among these, the virtuous person may think, is justice. Just as the deontologist may say we have a duty of justice to FastFresh employees not to use the AI system, the virtue ethicist may claim that a truly virtuous person would regard this system as inflicting a substantial injustice on those workers. It is, they may claim, grossly unfair. Moreover, the virtuous person may *feel* pained at the substantial harm inflicted on workers for the sake of profit and convenience. The virtuous person may also respond with *compassion* and a degree of anger towards the company. They may perceive that seeking to benefit shoppers and shareholders through this technology may superficially seem benevolent, but that it is not really benevolent. On the contrary, they might feel that aiming at producing profit and convenience is to miss the target or the mean with respect to the virtue of benevolence: It is instead to show a kind of excessive and misdirected altruism. A virtuous person may feel that a company which seeks by this mechanism to increase its profits is not acting in accordance with virtue but rather displays the vices of greed and callousness.

## The Care Ethicist Says ...

Finally, let's try our hand at briefly working through care ethics[6] for this case study. Again, it seems possible that a care ethicist will also be critical of management's use of the AI system. But they would nonetheless have their own reasons, and they may emphasise different aspects of the case to, say, some virtue ethicists. A care ethicist may concede that the managers of FastFresh have some responsibility to the shareholders they made a commitment to (and have legal duties towards). Nor need the care ethicist deny that managers may sometimes act in their own interests. All the moral theories, including care ethics, sometimes consider the moral agent's own interests. Care ethics recognises the importance of *self-care*, which also enables people to care more effectively for others. Most managers will have friends and family they care for deeply. Such relationships may give their lives purpose and meaning. For such reasons, care ethicists may say that seeking improvements in productivity and maintaining company competitiveness can be morally good.

That said, managers also have responsibilities for their employees. It may sometimes seem that the employer–employee relation is purely one of competition and rational self-interest. But the care ethicist may instead claim that there are or should be elements of care and mutual concern in this relationship. For example, managers cannot simply say to their workers: "*We agree that we will pay you the minimum wage and honour our legal contract. There is certainly no chance we will enslave you! Even so, to us you are essentially anonymous individuals, rather like robots that, lacking their own value and dignity, exist to serve the company*".

On top of that, warehouse workers are not usually well paid and tend to have relatively little social power. It is entirely possible that some employees already suffer disadvantage and discrimination. For example, employees in any industry, including this one, can have various physical and mental disabilities. Some employees are older and less physically or mentally capable. Some employees who are young and inexperienced may struggle with the challenges of entering the workplace. Some have difficulties finding work and would be in trouble if they were fired. Losing a job because you are slower than other workers could also undermine self-esteem; and so on.

A technology that constantly surveils, scrutinises, and ranks employees, putting them at continual risk of unemployment, has the potential to affect individuals very seriously, and in some cases may worsen existing disadvantage. FastFresh's AI system, then, hardly seems consistent with caring relations. The modest gains in productivity, the care ethicist may conclude, do not justify allowing people under one's care to suffer these mental and physical blows.

## Moral Theories Revisited

In the previous chapter, we introduced you to four mainstream ethical theories in western moral philosophy: Utilitarianism, Deontology, Virtue Ethics, and Care Ethics. We have just seen how these theories could be applied on a practical example.

Unless you have already sworn allegiance to one theory over the others, you might be quite uncertain about what the best or right theory is. Perhaps you are feeling confused or even irritated, because you think that the theories all have strengths and weaknesses and that none stands out as unproblematically true. The first thing to say here is that *it is totally OK*. You do not need to throw your hat in for a specific theory – at least not immediately, and perhaps not ever. In our opinion, there is no compulsion to become a card-carrying utilitarian, virtue ethicist, deontologist, or care ethicist.

Of course, you can certainly adopt a particular theory if you wish. And many people are more strongly attracted to one theory than to others. But even when we are personally undecided, the exercise of working through how a moral theory (even one we don't fully subscribe to) would or could apply to a moral issue or dilemma may be useful. For that exercise can shed much needed light on murky or difficult situations. Although the theories are strictly incompatible with each other, each basic framework may offer insights into morality. After all, consequences, duties, virtues, and care all seem to be very important features of ethics. These concepts are surely helpful in understanding and answering moral questions, and the fact that very thoughtful people have identified these ideas as central features of morality means that we may confidently take them seriously.

That said, we may note that some moral philosophers themselves do not personally adopt, or even have a preference for, a particular moral theory over the others. More strongly, some philosophers altogether oppose moral theorising. That might seem strange. However, philosophical critics of theory may argue, for example, that moral theories are too rigid. Critics may contend that the highly general criteria of rightness and value in moral theories fail to do justice to the complexity of our moral lives. For example, *moral 'particularists'* claim that basic moral duties or principles do not necessarily have broad and general application. Rather, they propose that we must work out what is right and wrong by immersion in particular cases and through our responsiveness and sensitivity to detailed situations. We may then recognise, the moral particularist may say, that no theory can dictate what is right, or accommodate all our considered moral responses.

Even so, such critics of theory may still agree that moral theories and their identification of various rules, duties, dispositions, and moral relations can lead to worthy ethical insights. Recognising that ethical theory is far from useless while still being controversial, some philosophers have proposed alternative ways of approaching, discussing, and debating thorny ethical issues. The approach we discuss next is called *principlism*. Perhaps even more than ethical theories, this approach has achieved a high profile in digital ethics.

## On Principlism

What is **principlism**? Most obviously, it is a set of ethical principles. Having these principles is a bit like having moral directions or guides to action. The four principles of principlism are **beneficence**, **nonmaleficence**, **respect for autonomy**, and **justice**.[7]

### Why Principlism?

Before we explain them, let's first ask why we might want to consider principlism at all. We can look upon principlism as an attempt to distil insights from moral theory and philosophy. The aim of this distillation is to produce a guiding framework that different moral theorists can at least to some extent agree on, notwithstanding their significant theoretical differences.

Principlism attempts this task by presenting a small set of *mid-level principles*. These action-guides are 'mid-level' principles in the sense that they are not basic or ultimate determinants of rightness or value. In this respect, the mid-level principles differ from, say, Ross's prima facie duties (which they resemble), Kant's categorical imperative, and the principle of utility, which all claim to be the singular root of ethically good behaviour. Yet the principles in principlism, like moral theories, also have some generality, distinguishing them from lower-level moral rules. For instance, the rule "do not use people's digital data without their consent", is, though possibly correct, extremely specific for a moral principle. Low-level directives like this cannot helpfully address the wide assortment of moral questions that digital technologies pose.

Principlism offers only several principles, partly because a few principles are arguably all we need, and partly because a large set of principles would be less manageable and less useful. As with Ross's deontology (Chapter 4), *these four principles are prima facie*. That is, they provide stable moral reasons for actions across the areas of human life, but they can sometimes be overridden, depending on circumstances. All four principles are in a sense equal: None has automatic priority over the others. Accordingly, the prima facie principles are non-hierarchical, and our judgement is required in their application and when they conflict.

A moral theorist might be happy to use the mid-level principles in moral discussion and debate about digital technology. Unsurprisingly, each moral theorist will ultimately justify the principles by reference to their theory. Yet so long as moral theorists can agree on the role of select mid-level principles, they can partly – though not always and not entirely – put aside their deeper moral differences. In this way, moral theorists can try to make some progress in gaining agreement with others about various contentious, challenging, or novel ethical matters. Finally, those who do not adopt a moral theory can still deploy principlism to evaluate digital technologies.

Principlism in digital ethics is borrowed from medical ethics. Its appearance in healthcare, in the late 1970s work of Thomas L. Beauchamp and James F. Childress,[8] followed appalling wrongdoing by medical practitioners. This included the unjust, harmful, and racist exploitation of African Americans with syphilis in the Tuskegee experiments which we briefly hinted at in Chapter 1.[9] Principlism also arose because of general ethical confusion about how doctors should behave. Analogously, principlism in digital ethics followed scandals involving big tech, and ethical uncertainty about the impact of powerful new technologies in the digital age.

Let's now explain the four prima facie principles and show how they echo elements of moral theory.

## Beneficence

The principle of *beneficence* enjoins us to do good for others and promote their wellbeing, such as their psychological, physical, and social interests. As you might recall, Ross's deontology too has a prima facie duty of beneficence. However, beneficence in principlism also echoes utilitarianism's goal of promoting happiness, virtue ethics' virtue of benevolence, and care ethics' responsibility of caring for others and their needs.

### Nonmaleficence

The next principle, *nonmaleficence*, also concerns wellbeing, but it focuses on avoiding doing harm. Again, we can detect resonances with Ross's duty of nonmaleficence, the principle of utility, the character trait of benevolence, and relations that avoid causing harm to those being cared for.

### Respect for Autonomy

The third principle of *respect for autonomy* tells us to recognise and act in certain ways towards individuals who have autonomy. Kant's form of autonomy required an ability to act on the moral law. However, a broader (and less intellectualist) conception of autonomy is the ability to possess values and plans and to direct the course of one's life. Respecting autonomy can require giving individuals the freedom, and the necessary support, to express themselves and pursue preferred goals. Sometimes individuals will need or want the help of others in making decisions.

Respecting autonomy can also require ensuring individuals understand enough about technologies' risks and benefits, and that they have the opportunity to provide informed consent to treatment that affects their values and plans. For example, not adequately giving people clear information about private data collection or making it onerous to opt out of online profiling both affect autonomy. Respect for autonomy may be considered important by utilitarians because it tends to increase happiness, by virtue ethicists because it is required by benevolence and justice, and by care ethicists because it deepens relationships while meeting our vital need to govern our own lives.

### Justice

The fourth principle of *justice* requires treating people as they deserve to be treated, without prejudice or wrongful discrimination. This can involve the fair allocation of harms and benefits, as well as compensating victims who have been wrongly harmed. Deontology, virtue ethics, and care ethics tend to intrinsically value fairness. Because utilitarianism prioritises overall happiness, it is often criticised for requiring inequitable distribution of harms and benefits. For example, utility might be maximised by harming one person to make several others better off. In practice, however, utilitarians may agree that fairness often maximises utility. For example, they may argue that inequitable societies have much worse welfare outcomes than societies with a smaller gap between rich and poor. Thus, even utilitarians may sometimes use principlism's justice principle.

### Potential Objections to Principlism

An objection to 'digital principlism' relates to its medical ethics origins. In healthcare, operationalising the four classic principles, though requiring careful thought, is relatively *limited* in scope. In contrast, digital technologies, as we noted, are enormously *varied* in nature and usage. For example, data, software, robots, and AI can be found in industry, government, medicine, law, policing, banking, sport, education, farming, and much more. The worry is that the four principles may not translate well beyond the confines of medicine. However, the four principles do seem to capture some general and important moral considerations that are broadly consonant with ethical theories. Also, scholars have found them useful for analysing digital ethics cases.

A second, different, objection is that digital principlism is unnecessary because it is just *obvious* when, say, tech companies and digital practitioners are behaving badly. The same charge has been levelled at medical principlism: *"It is all too obvious"*, the critic of principlism may argue, *"that doctors ought not to harm their patients, ignore their wishes, and so on"*. The reply to this charge is that medico-ethical cases are not always straightforward, and indeed that hard dilemmas can arise. Furthermore, due to human nature and the demands of the job, doctors may sometimes be tempted to ignore or forget basic principles. Both these points remain true for digital technologies and their creation. Indeed, it has often been said that the Silicon Valley mindset and the 'tech-bro' culture[10] of moving fast and breaking things[11] or innovating at all costs is an excellent environment for nurturing ethical flippancy and forgetfulness.

## Principlism in Action: Outsourcing Digital Content Review

To glimpse how these principles could work, we will consider another case study. As before, we can only sketch out one possible, simplified ethical scenario for analysis:

> In recent years, prominent US tech corporations have outsourced to other companies and countries the work of reviewing platforms and AI products for offensive and illegal content. The business activity of outsourcing this difficult work to low wage nations like Kenya and India is controversial. It has been condemned as exploiting foreigners in the Global South. These workers had to identify extreme content such as child abuse, dismemberment, suicide, and animal cruelty. However, the work gave unemployed people a wage, albeit significantly less than an American would receive for the same work.[12]

The question, is, therefore:

> Were these actions of US tech companies ethical?

We can begin our argument via the four principles of beneficence, nonmaleficence, respect for autonomy, and justice.

The outsourcing of content review to the Global South arguably did good by giving poor and desperate people much-needed employment. In addition, removing obscene, graphic content from platforms protected millions of technology users, including children, from viewing extremely upsetting material. These actions might perhaps also deter the future creation of some such content. In these respects, then, the principle of beneficence seems to support the companies' actions. (Incidentally, the example of providing employment shows that digital technology can have benefits beyond the actual tasks it performs. Of course, one of the big worries about technology such as AI and robots is that it could cause mass employment.)

But now consider the other three principles. According to reports, the contracted workers had to view horrendous photos, text, or video day after day for hours at a time. This had a profoundly traumatic and long-lasting psychological effect on many individuals. Some had to stop work and marriages broke up. This looks like a serious transgression of the principle of nonmaleficence, which tells us to not cause harm.

It might nonetheless be argued that the companies respected the autonomy of the workers, who were not forced to enlist. Surely, it was their choice to take on risky work? However, these individuals who took on this difficult work were often in desperate circumstances, which could have affected their ability to make freely chosen, autonomous choices to accept such potentially awful consequences. Moreover, many workers were apparently not properly informed of the psychological and social risks of repeated exposure to horrific content. If so, then their autonomy was doubly disrespected.

At first blush, the principle of justice may also seem to support the tech companies' actions. Giving work opportunities to people in severely disadvantaged countries appears to more fairly distribute benefits to those in most need. However, for many of these people, the work was at least as harmful as it was beneficial. Adequate counselling for psychological injury was often not provided. Furthermore, although the workers were paid, their wage of $1.50–2 an hour was extremely low compared to pay in the Global North. In conclusion, while the principle of beneficence could be used to defend the companies' behaviour, the other principles arguably speak forcefully against it.

Thus, beneficence, in this particular case at least, may be *overridden* by the other prima facie principles of nonmaleficence, respect for autonomy, and justice. This lesson is worth remembering. Sometimes technologies will indeed bring about significant improvements in our lives and so seem to satisfy the principle of beneficence. But that is not always the case. Furthermore, the other principles can sometimes (though not always) outweigh the apparent beneficence of making and introducing new technologies. (This discussion of competing and overriding principles will be revisited in Chapter 6).

## Principlism Revisited: Additional Digital Ethics Principles

Some philosophers of AI have suggested adding a fifth principle to digital principlism: **explicability**.[13] The principle of explicability seeks to ensure AI systems are intelligible or understandable to users and that designers and operators are accountable for their risks (see below). While this is important, we *haven't* included explicability as a mid-level principle. Despite AI's current prominence, digital ethics covers a wider array of digital technologies where explicability may be less salient.

Moreover, explicability appears *derivable* from the other four principles, which are thus more morally basic (though still as we said 'mid-level'). For comparison, the founders of medical principlism, Beauchamp and Childress, argued that duties of, say, patient confidentiality can be *derived* from mid-level principles, such as respect for autonomy and nonmaleficence, because breaching confidentiality is usually counter to patients' values and wishes and is distressing.

Similarly, there are various lower-level and derivable rules or principles that are important in the digital arena. Some of these concepts are less technical (e.g., inclusiveness), while some more obviously arise from the specific natures of digital technologies (e.g., algorithmic transparency). We discuss some additional digital ethics principles now, in no particular order.

Note that some principles are touched on in other chapters – some of which directly correspond to ethical issues as we categorised in Chapter 3. Also, note that the following is not an exhaustive list, though it draws on reviews of commonly identified principles concerning technology – often AI.

### Safety

Digital technologies are becoming increasing pervasive and impactful. We clearly want digital technologies, in healthcare, domestic environments, workplaces, and many other domains, to be safe and non-harmful for operators and users. There is a clear link here to the principle of nonmaleficence. **Safety** may require adequate testing prior to and after deployment of the technology, perhaps throughout the product's lifecycle from inception to retirement. If a technology cannot be made perfectly risk-free, a principle of safety might, nonetheless, direct practitioners to make them as safe as the circumstances morally require. It is worth noting, however, that some technologies, like military technology, are expressly designed to cause harm.

## Privacy

**Privacy** is a complex concept,[14] which we briefly highlighted in Chapters 2 and 3. It can, for example, refer to individuals having seclusion from observation or control over personal information, such as information about one's body, health, beliefs, and relationships. Privacy can refer to others having potential access to such information, such as when data is collected by a platform or app and stored, or to a human being actually viewing that information.

Digital technologies are increasingly able to capture personal details, re-identify individuals from anonymised data, and make further inferences (if often inaccurately) about beliefs and personalities from a limited dataset. The commodification of personal data by tech corporations has been dubbed *'surveillance capitalism'*.[15] Data security is increasingly a major consideration, and hacking and fraudulent impersonation to steal personal data is worryingly common. As alluded to above, a duty to protect privacy may follow from mid-level principles, such as nonmaleficence and respect for autonomy. A duty of confidentiality arises from special relationships or promises, such as when a tech company pledges to users not to sell their data to third parties.

## Transparency

A principle of **transparency**, of being open and perceivable, invokes a range of other concepts such as interpretability, **explainability**, and disclosure. We have provided a birds' eye view of transparency and its related concepts in Chapter 3, and in this section, we will expand more on its role as a digital ethics principle.

**Interpretable** intelligent systems are those whose outputs, perhaps in the form of decisions or inferences, are understandable. This may mean that the basis or reasons for the machine's output are intelligible to a human being. Many outputs from deep learning systems are not interpretable in this way, since they are too mathematically complicated to be humanly understood; take, for example, large language models nowadays which have billions – or even trillions – of variables![16] This is why deep neural networks are often labelled opaque algorithms or *black boxes*.

Explainability is the ability to provide various kinds of explanations of the outputs of such systems, even when they are opaque. Explainability is now its own subfield of AI research.

**Disclosure** may refer to providing information about the nature of a system, such as data provenance, machine training methods, and possible biases. Often, for propriety reasons, companies do not disclose this information. Sometimes that may conflict with principles such as respect for autonomy and justice. However, disclosure of information or data can sometimes also cause harm, such as unwanted loss of privacy and nefarious gaming of systems that lead to maleficent outcomes.

Transparency, as a concept, has its nuances. Consider a common example in our Preface: You are curious as to why a social media platform is showing you a particular advertisement for a product X. Some modern platforms will be 'transparent' in providing *some* justification (e.g., *"you were using the app in a particular venue which has a high interest in X"*), but you are still left unconvinced. Researchers further distinguish different 'levels' or *'degrees' of transparency*. For instance, Philip Lorenz-Spreen's team contrast between 'nominal' vs 'effective' transparency[17]: The latter helps users understand, while the former tends towards the minimum 'lip service'.

## Liberty

**Liberty** may be considered good because it promotes wellbeing and respects autonomy. There are various forms of liberty or freedom, including freedom of expression, freedom to information, and freedom from digital manipulation.[18] Digital technologies are increasingly adept at distorting the information we receive and pushing and nudging us in various directions, such as to purchase products or vote for certain candidates. A digital principle of liberty reminds us to consider the impacts of technology on our important freedoms.

## Trustworthiness

In many cases, we want to be able to trust that technologies do what they say they will or what they are supposed to do. *Justified* trust will require that those technologies are **trustworthy**.[19] For example, we may want to trust that an AI system does not create undue harm and that it provides certain benefits, respects users' autonomy, and is not unfair. We may also need the technology to be consistently accurate or robust rather than brittle across different applications. That can sometimes be a problem when, for example, a machine learning algorithm is applied, or transferred, to new sets of data in the real world. When the aforementioned risks are present, it may sometimes be morally necessary to alert users and deployers so that they rightly distrust such technology. This may imply not implementing the technology or else exercising due caution, to minimise harm and promote benefit, fairness, and autonomy.

## Dignity

**Dignity** is another complex ethical concept, though it is less discussed in digital contexts. For Kant, dignity concerned the special intrinsic value of autonomous moral agents, who warrant special respect. All moral agents have equal dignity regardless of their origins, race, background, level of virtue, and so on. Yet while Kant restricted dignity to moral agents, and thus to typically functioning human adults, most of us would be loath to specifically deny dignity to human individuals who lack full moral agency. This includes severely intellectually impaired humans and people with advanced dementia. Lately, philosophers have also argued that nonhuman animals, especially sentient and sapient animals, have a kind of dignity and significant intrinsic value. Indeed, some digital ethicists say we should take more notice of the impact of digital technologies on nonhuman beings.

Human dignity is often seen to underpin human moral value and also universal human rights, such as legal or ethical rights to life, liberty, a fair hearing, and freedom from torture and servitude.[20] Protecting people from human rights impacts of technology may be regarded as a priority. Tech companies that apparently show insufficient attention to human rights have sometimes been denounced. For example, Meta was heavily criticised for failing to stop human rights abuses in several countries.

In a sense, dignity (unlike the other more specific principles we list here) seems better thought of as even more fundamental than the mid-level principles, rather than being an implication or extension of them. That is, perhaps we have obligations of beneficence, justice and the like *because* humans and perhaps some animals have dignity.[21] On the other hand, there is another form of dignity that we could arguably link to, for instance, principles of justice, respect for autonomy, and/or nonmaleficence. Respect for this form of dignity concerns the avoidance of *degrading* or *demeaning* treatment. Digital examples of such treatment are humiliating cyberbullying and pornographic deepfakes.

### Inclusiveness

Minority, oppressed, and often forgotten groups may sometimes be especially at risk of harm or the failure to benefit from digital technologies. These groups could include the physically and intellectually disabled and very old or very young people. A principle of **inclusiveness** requires that digital technologies be accessible and useable for all these groups. To achieve that, it is often necessary to incorporate group representatives in an inclusive design process (a point we will touch on in Chapter 7). Such inclusiveness could also be seen as a requirement of justice, such as in the Maguire vs SOCOG case in Chapter 2.

### Accountability

The principle of **accountability** requires that relevant actors, such as designers, deployers, organisations, and governments, take responsibility for ethically relevant implications of technologies.[22] This could include enacting measures to ensure technology operates fairly, safely, transparently, and so on across its lifecycle. We discuss some of these measures in another chapter. It could also require ethical processes for the implementation of technologies, such as democratic and procedurally fair mechanisms that give stakeholders, including vulnerable parties, a voice. It could further require contestability, such as allowing individuals subject to automated decision-making to question the decision, ask for a review, or seek another decision by a machine or a human.

Technologies such as some AI-automated systems that act more-or-less independently may seem to create a *responsibility or accountability 'gap'*. For example, it may be unclear who has responsibility amongst, say, designers, deployers, and authorities for an automated decision system that shows unfair bias against some individuals or proves to be inaccurate. In this case, work may need to be done to assign clear responsibility to one or more parties. In addition, having humans in or on the loop with the autonomous system to prevent bad outcomes may be ethically required.

### Solidarity

Because digital technologies are increasingly powerful and disruptive, people are recognising that preventing harmful and unjust uses can require coordinated action (Chapter 7) by individuals who have common goals and vulnerabilities. **Solidarity** may be seen as an ethical principle of joining together and taking a stand against technologies or tech companies that, for example, unjustifiably seek large profits at the expense of workers, artists, disadvantaged people, Indigenous groups, and so on.

### Sustainability

**Sustainability** has been advanced as an important digital ethics principle especially because technologies such as data centres and large AI systems are energy-intensive, can have a high carbon footprint, generate waste, require mining for rare minerals, and consume copious amounts of water[23] (see also Chapter 3). Digital technology has the potential to cause both local harm and planetary harm. A principle of sustainability might direct us to mitigate negative environmental impacts for the sake of humans, including those most vulnerable to climate change, like people in the Global South. It may require us to protect ecosystems, biodiversity, and individual animals who, like humans, suffer and die, often in very large numbers, from global heating and environmental disturbance.

* * *

As we indicated, this is not an exhaustive list of additional and/or more specific principles. Note also that some ethical issues are not necessarily well covered or informed by these other principles. For example, there has lately been acute concern about the way that creators of large language models, image generators, and other AI modalities have used the works of artists and other internet users to train algorithms without permission or payment. (This has sometimes resulted in acts of solidarity, such as strikes against replacing actors with AI imitators, as seen in Hollywood[24]). If we recall here the four principles of digital principlism, we will see that principles such as justice and respect for autonomy may be used to criticise this behaviour.

At the same time, however, we should remember that some principles may point in other directions. For example, we might need to also consider the benefits that such generative AI products create. Furthermore, we should emphasise that there are different possible ways of interpreting the principles. For example, some creators of those AI products claim that using paintings, photos, and other material from the internet to train image generators does not infringe copyright and are not unfair (in contrast to claims seen in, say, Chapter 3). That is because, they argue, the AI does not merely duplicate those images but only imitates elements of their style, as a human artist might.

Irrespective of where we stand on this debate, our point here is that we cannot assume that the interpretation and application of principles of justice, nonmaleficence, beneficence, and respect for autonomy will always be obvious and uncontroversial. Essential to the principlist approach, then, is carefully thinking through how those principles apply in different scenarios, as well as what to do when prima facie principles conflict.

Clashes are inevitable, no matter if they belong to the four core principles, or the additional ones. We will revisit what happens when principles clash in Chapter 6. Before we conclude this chapter, it is vital that we recognise the plurality of ethical traditions globally, and their increasing importance in the field of digital ethics.

## Moving Beyond the *Canon*

As we have pointed out before, the majority of the ethical thought in digital ethics rests upon Western perspectives, in the so-called Western philosophical canon.[25] However, there is a plurality of ethical perspectives from non-Western traditions.

And yes, these perspectives are very valuable in understanding concerns surrounding technology and its impact on us and our surroundings.

We highlight a few other complementary ethical perspectives here.[26]

### Ubuntu Philosophy

Some digital technologists may have heard of the term 'Ubuntu' from its namesake Linux operating system distribution. **Ubuntu philosophy** goes beyond that, as a rich tradition originating from South Africa.

Ubuntu – from the Bantu phrase *umuntu ngumuntu ngabantu* (widely interpreted as 'I am because we are') – emphasises concepts such as interconnectedness, togetherness, and connection, with respect to one's community and people around them. Famously, Ubuntu was adopted by Archbishop

Desmond Tutu in implementing the post-apartheid Truth and Reconciliation Commission.[27]

Many emerging studies use Ubuntu philosophy in their discussions on topics as diverse as digital technologies and entrepreneurship,[28] good IT governance,[29] as well as evaluating the deployment of software created by Western organisations, which may not necessarily reflect cultures in the Global South.[30]

### Buddhist Philosophy

Most people have heard the term 'mindfulness', often in relation to stress reduction and positive psychology. Yet the practice of mindfulness meditation is adopted from **Buddhist philosophy** advanced by, for instance, teachers like Thích Nhất Hạnh. Based on current research,[31] mindfulness has positive impacts on health and well-being; one doesn't have to be religious to reap its benefits. And mindfulness is a useful practice given our daily use of technologies such as our mobile phones and tablets. Mindfulness is just one of the many ways ideas from Buddhist philosophy can help us in today's lure of the 'attention economy'.[32]

Beyond mindfulness (or *Sati*), there is increasing research to show that these principles – including *Ahimsa* (compassion and non-violence towards humans and other living beings) and *Pratītyasamutpāda* (interdependence) – can be valuable in guiding our use of technology. To borrow from the philosopher Soraj Hongladarom, "it is a universal truth that everyone wants to enjoy happiness and avoid suffering",[33] and Buddhist philosophy can offer useful insight into our overall engagement with the digital in the everyday.

### Other Schools of Philosophy

Many other schools of philosophy as well as religious traditions from around the world have been studied by digital ethics researchers and philosophers.[34] For example, the Japanese concept of *ikigai* – roughly translated to 'meaning to life' or 'sense of purpose' – can be used to study our relationship to automation and robotics.[35] Research in Islamic philosophy notes that the concept of *masala* (translated as 'good') has the potential of being 'a normative guide for AI's ethical evaluation'.[36] Perspectives from Indigenous knowledge, such as Māori and Australian Aboriginal and Torres Strait Islander people, could also provide us with valuable lessons for the design and evaluation of AI systems.[37]

* * *

This chapter has hopefully provided you with an opportunity to exercise your 'ethical thinking muscles' on a couple of case studies that are plausibly realistic in today's digital environment. We have also visited yet another moral approach, principlism, which has its roots in applied medical ethics, as well as briefly noting examples of non-Western philosophies that are fast gaining traction in digital ethics.

However, ethics is not the be all and end all – not the proverbial 'silver bullet' or 'panacea' for a better world in the digital era. In the next chapter, we will discover the limitations of (digital) ethics itself, but also, forewarn you on how ethics can be misused.

## Notes

1. See, for example, write-ups in the popular press:
   Kantor, Sundaram, Aufrichtig & Taylor (2022, August 15).
   Murty & Karanth (2022, October 27).
2. de Lazari-Radek & Singer (2017)
3. Simpson (n.d.).
4. Kant & Ellington (1993).
5. Aristotle & Ross (2009).
6. As seen in Chapter 4. See also: Held (2006).
7. Beauchamp & DeGrazia (2004).
8. First published in the late 1970s, *The Principles of Biomedical Ethics* has been republished in subsequent editions; we base our discussion on Beauchamp & Childress (1979).
9. CDC (2025, February 13).
10. Griffith (2022, September 24).
11. Taneja (2019, January 23).
12. Perrigo (2023, January 18).
13. Floridi & Cowls (2019).
14. Solove (2002).
15. Zuboff (2019).
16. Giattino, Mathieu, Samborska & Roser (2025, March 12).
17. Lorenz-Spreen et al. (2021).
18. Mill (1989).
19. Jacovi, Marasović, Miller & Goldberg (2021).
20. Rosen (2012).
21. Coghlan (2025).
22. Novelli, Taddeo, & Floridi (2024).
23. van Wynsberghe (2021).
24. SAG-AFTRA (2023, August 18).
25. For further reading of what the 'canon' is, and scholarly attempts to go 'beyond' the canon, see, e.g., Van Norden (2017) and Park (2013).
26. Special thanks to Raffaele Ciriello for bringing this, and related literature, to the authors' attention. Ciriello is also a collaborator on an upcoming manuscript on Compassion and Digital Innovation, which also draws upon these philosophies.
27. Ifejika (2006, September 28).
28. Abubakre, Faik, & Mkansi (2021).

29. Zubler, Plattfaut, & Niehaves (2025).
30. Kroeze (2024).
31. See, for example, the Contemplative Studies Centre at The University of Melbourne: https://psychologicalsciences.unimelb.edu.au/CSC.
32. Bombaerts et al. (2023).
33. Hongladarom (2020).
34. Ziesche (n.d.), Eddebo (2021), and Dennis et al. (2024).
35. Ziesche (n.d.).
36. Elmahjub (2023).
37. Walter & Kukutai (2018) and Munn (2024).

## References

Abubakre, M., Faik, I., & Mkansi, M. (2021). Digital entrepreneurship and indigenous value systems: An Ubuntu perspective. Information Systems Journal, 31(6), 838–862. https://doi.org/10.1111/isj.12343

Aristotle, & Ross, W. D. (2009). The Nicomachean Ethics (W. D. Ross & L. Brown, Trans.). Oxford University Press.

Beauchamp, T. L., & Childress, J. F. (1979). Principles of Biomedical Ethics (1st ed.). Oxford University Press.

Beauchamp, T. L., & DeGrazia, D. (2004). Principles and principlism. In Handbook of Bioethics: Taking Stock of the Field from a Philosophical Perspective (1st ed.). (ed. 2004). Imprint: Springer. https://doi.org/10.1007/1-4020-2127-5

Bombaerts, G., Anderson, J., Dennis, M., Gerola, A., Frank, L., Hannes, T., Hopster, J., Marin, L., & Spahn, A. (2023). Attention as practice: Buddhist ethics responses to persuasive technologies. Global Philosophy, 33(2), 25. https://doi.org/10.1007/s10516-023-09680-4

CDC. (2025, February 13). About The Untreated Syphilis Study at Tuskegee. The U.S. Public Health Service Untreated Syphilis Study at Tuskegee. https://www.cdc.gov/tuskegee/about/index.html

Coghlan, S. (2025). Nonhuman animal dignity. Philosophy Compass, 20(4), e70035. https://doi.org/10.1111/phc3.70035

de Lazari-Radek, K., & Singer, P. (2017). Utilitarianism: A Very Short Introduction. Oxford University Press.

Dennis, M., Frank, L. E., Herbener, A. B., Klincewicz, M., Damholdt, M. F., Puzio, A., Bassil, K., Stone, J., Schneidenbach, P., Das, S., Thomas, E., & Rawsthorne, M. (2024). Reconceptualizing the Ethical Guidelines for Mental Health Apps: Values from Feminism, Disability Studies, and Intercultural Ethics. IEEE Xplore, 1–33.

Eddebo, D. J. (2021). The faustian machine and the chrome Lotus: On the diversity of perspectives on the metaphysics of artificial intelligence with a particular focus on the contributions of traditional non-Western thought. New Techno Humanities, 1(1), 100001. https://doi.org/10.1016/j.techum.2021.100001

Elmahjub, E. (2023). Artificial intelligence (AI) in Islamic ethics: Towards pluralist ethical benchmarking for AI. Philosophy & Technology, 36(4), 73. https://doi.org/10.1007/s13347-023-00668-x

Floridi, L., & Cowls, J. (2019). A unified framework of five principles for AI in society. Harvard Data Science Review, 1(1). https://doi.org/10.1162/99608f92.8cd550d1

Giattino, C., Mathieu, E., Samborska, V., & Roser, M. (2025, March 12). Data Page: Parameters in Notable Artificial Intelligence Systems. Our World in Data. https://ourworldindata.org/grapher/artificial-intelligence-parameter-count

Griffith, E. (2022, September 24). Silicon Valley Slides Back Into 'Bro' Culture. The New York Times. https://www.nytimes.com/2022/09/24/technology/silicon-valley-slides-back-into-bro-culture.html

Held (2006). The Ethics of Care: Personal, Political, and Global. Oxford University Press.

Hongladarom, S. (2020). The Ethics of AI and Robotics: A Buddhist Viewpoint. Lexington Books.

Ifejika, N. (2006, September 28). What Does Ubuntu Really Mean? The Guardian. https://www.theguardian.com/theguardian/2006/sep/29/features11.g2

Jacovi, A., Marasović, A., Miller, T., & Goldberg, Y. (2021). Formalizing trust in artificial intelligence: Prerequisites, causes and goals of human trust in AI. Proceedings of the 2021 ACM Conference on Fairness, Accountability, and Transparency, 624–635. https://doi.org/10.1145/3442188.3445923

Kant, I., & Ellington, J. W. (1993). Grounding for the Metaphysics of Morals; With, On a Supposed Right to Lie Because of Philanthropic Concerns (3rd ed.). Hackett Pub. Co.

Kantor, J., Sundaram, A., Aufrichtig, A., & Taylor, R. (2022, August 15). The Rise of the Worker Productivity Score. The New York Times. https://www.nytimes.com/interactive/2022/08/14/business/worker-productivity-tracking.html

Kroeze, J. H. (2024). Digital coloniality. An Ubuntu perspective. Scandinavian Journal of Information Systems, 36(2). https://aisel.aisnet.org/sjis/vol36/iss2/6

Lorenz-Spreen, P., Geers, M., Pachur, T., Hertwig, R., Lewandowsky, S., & Herzog, S. M. (2021). Boosting people's ability to detect microtargeted advertising. Scientific Reports, 11(1), 15541. https://doi.org/10.1038/s41598-021-94796-z

Mill, J. S. (1989). JS Mill: 'On Liberty' and Other Writings. Cambridge University Press.

Munn, L. (2024). The five tests: Designing and evaluating AI according to indigenous Māori principles. AI & SOCIETY, 39(4), 1673–1681. https://doi.org/10.1007/s00146-023-01636-x

Murty, R. N., & Karanth, S. (2022, October 27). Monitoring Individual Employees Isn't the Way to Boost Productivity. Harvard Business Review. https://hbr.org/2022/10/monitoring-individual-employees-isnt-the-way-to-boost-productivity

Novelli, C., Taddeo, M., & Floridi, L. (2024). Accountability in artificial intelligence: What it is and how it works. AI & SOCIETY, 39(4), 1871–1882. https://doi.org/10.1007/s00146-023-01635-y

Park, P. K. J. (2013). Africa, Asia, and the History of Philosophy: Racism in the Formation of the Philosophical Canon, 1780–1830. State University of New York Press.

Perrigo, B. (2023, January 18). Exclusive: OpenAI Used Kenyan Workers on Less Than $2 Per Hour to Make ChatGPT Less Toxic. Time. https://time.com/6247678/openai-chatgpt-kenya-workers/

Rosen, M. (2012). Dignity: Its History and Meaning. Harvard University Press.

SAG-AFTRA. (2023, August 18). Support Strong for Striking Entertainment Industry Workers, Poll Finds. SAG-AFTRA. https://www.sagaftra.org/support-strong-striking-entertainment-industry-workers-poll-finds

Simpson, D. L. (n.d.). William David Ross (1877–1971) [Internet Encyclopedia of Philosophy]. Internet Encyclopedia of Philosophy. Retrieved March 25, 2025, from https://iep.utm.edu/ross-wd/

Solove, D. J. (2002). Conceptualizing privacy. California Law Review, 90, 1087.

Taneja, H. (2019, January 23). The Era of "Move Fast and Break Things" Is Over. Harvard Business Review. https://hbr.org/2019/01/the-era-of-move-fast-and-break-things-is-over

Van Norden, B. W. (2017). Taking Back Philosophy: A Multicultural Manifesto. Columbia University Press.

van Wynsberghe, A. (2021). Sustainable AI: AI for sustainability and the sustainability of AI. AI and Ethics, 1(3), 213–218. https://doi.org/10.1007/s43681-021-00043-6

Walter, M., & Kukutai, T. (2018). Artificial Intelligence and Indigenous Data Sovereignty. Input paper for the Horizon Scanning Project "The Effective and Ethical Development of Artificial Intelligence: An Opportunity to Improve Our Wellbeing." Australian Council of Learned Academies. https://acola.org/wp-content/uploads/2019/07/acola-ai-input-paper_indigenous-data-sovereignty_walter-kukutai.pdf

Ziesche, S. (n.d.). Stimuli from Selected Non-Western Approaches to AI Ethics. Retrieved March 26, 2025, from https://philarchive.org/rec/ZIESFS

Zubler, M.-E., Plattfaut, R., & Niehaves, B. (2025). Decolonizing IT governance in international non-governmental organisations: An Ubuntu approach. Information Systems Journal, 35(1), 163–208. https://doi.org/10.1111/isj.12541

Zuboff, S. (2019). The Age of Surveillance Capitalism: The Fight for a Human Future at the New Frontier of Power. Profile Books.

# 6

# *No Silver Bullet*

## Principlism: A Promising 'Silver Bullet'?

Throughout this book, we have emphasised the importance of careful ethical thinking and consideration to recognise, avert, and diagnose problems that arise from the creation and use of technology.

In Chapter 4, we looked at the four ethical theories commonly used in (digital) ethics, namely utilitarianism, deontology, care ethics, and virtue ethics. In Chapter 5, taking a leaf from applied biomedical ethics, we considered the adaptation of ethical principles as a useful and relatively easy-to-understand tool to guide ethical evaluation and practice with respect to digital technologies.

In the spirit of the ethical approach of principlism, various organisations have produced ethics guidelines for the consideration, design, and use of digital technologies.[1] Often these include voluntary sets of principles that resemble the principles discussed in Chapter 5. For example, guidelines from organisations and bodies often include principles of transparency, accountability, privacy, and justice.[2]

Organisations that publish binding or non-binding principles and other ethical guidance include tech corporations,[3] governments,[4] NGOs,[5] academic groups,[6] and world bodies such as the UN[7] and OECD.[8] The recent interest in artificial intelligence (AI) has meant that many of these guidelines are about that technology. However, AI guidelines often have application to other or related digital technologies. There are also codes of ethics produced by professional organisations such as the IEEE[9] and ACM,[10] which members are supposed to follow or at least consider. Various organisations may create aspirational digital ethics guidelines to guide practitioners and managers, inform users, influence legislation, and vouch for the ethical integrity of an organisation.

## Four Issues with the 'Silver Bullet'

A key reason for creating and deploying these ethical principles is that they can help us to make and implement morally better and well justified technologies. However, there is no 'silver bullet' to the creation of ethical

DOI: 10.1201/9781003393511-6

technology. To explain this, let's consider a relatable analogy: Setting a new year's resolution.

> It is the end of December and time to make a plan for the coming year. *New year, new start*. Let's say you, a person who perhaps spends too much time working, come up with the admirable resolution to (a) do some proper exercise and (b) spend time every week doing something nice for yourself. Both these principles, you believe, will contribute to a better state of wellbeing and give a better work/life balance than working 7 days a week.

Playing devil's advocate, let's consider how these principles may go awry in practice. As we do this, we will also see issues with advancing and following ethical principles associated with principlism.

## Objection 1: Lack of Enforceability, Vagueness, and Cherry Picking

### Scenario 1: Quitting the gym and selective resolution-keeping

> Imagine that the year has ticked over and that you've started a gym membership. Going to the gym is, undoubtedly, an effort. After several attendances, you contemplate quitting the gym and giving up on part (a) of your resolution. After all, nobody needs to know, right? Nobody can force us to stick to our resolutions besides ourselves.
>
> For the other part (b) of the resolution, where we need to do something nice for ourselves, it is vague enough and easy enough to be achieved (you think), so you are comforted by the fact that you are at least sticking to one of your resolutions.

Thus, problems with our New Year's 'better work/life balance' principles include lack of enforceability and vagueness. Similar problems may apply to someone, or to an organisation, who adopts a set of digital ethics principles. If there are no mechanisms to make one accountable to the avowed ethical principles, and if some of the principles seem to be vague anyway, might we simply not bother with this or that principle? One can imagine, say, a tech company doing just that.

Critics of ethical principles for technologies also claim that they are too abstract and high level, and it is therefore hard for individuals or organisations to know how to make them concrete and operationalise them. Indeed, the criticism may go, individuals or organisations may choose to apply the ethical principles irresponsibly, cherry-picking as it suits them. After all, who is watching? What should we think of this sort of criticism of ethical principles?

In Chapter 5, we acknowledged that the mid-level principles of principlism and the additional ethical principles that we discussed must be interpreted and applied carefully. Corresponding justifications and arguments need to be given. (That may also sometimes include turning to moral theory for assistance.)

At the same time, the more abstract and general nature of mid-level principles, and even of various principles derived from them, are part of the very thing that makes these action-guides very useful and important. This, of course, need not stop more specific ethical recommendations being made.

As we say, ethical principles need to be applied thoughtfully and also sincerely. Clearly, we cannot force people to be thoughtful and sincere, just as we cannot force ourselves to comply with a New Year's resolution if we are hell-bent on finding excuses. Still, there may be other ways to encourage impartial ethical thinking and action.

For example, we could create ethics committees to debate difficult issues and argue about the application ethical principles. We might also seek to foster a reflective, honest, and courageous culture in an organisation to encourage shared, critical discussion. In these and other ways, we might seek help in interpreting ethics principles and in sincerely or genuinely working out how they apply to particular cases.

## Objection 2: Conflict

### Scenario 2: Trading off the gym

Assume, now, that the only gym in town moves to another town that is some distance away. Because it now takes so long to get there, you feel that you have to choose between going to the gym and relaxing and having fun on your one day off work. But *how* can you decide? You feel conflicted and confused.

In this case, the principle of doing proper exercise apparently conflicts with the principle of spending time on leisure activities. This is another common objection to principlism or to sets of ethical principles. Conflicting principles, as with new year's resolutions, can lead to bothersome uncertainty. In fact, we already know that it is the case with Beauchamp and Childress's four principles[11] (Chapter 5). To recap:

- The four principles – benevolence, nonmaleficence, justice, and autonomy – are prima facie not absolute.
- All four principles are in a sense equal and non-hierarchical.
- Judgement is required in their application and when they conflict.
- They can sometimes be overridden.

Consider an example. An organisation making an AI model that involves training on sensitive information (e.g., personal health information) might rightly consider privacy of the data as a critical ethical principle. However, privacy may conflict with an ethical principle of transparency, that is, with the moral need to disclose data used to train its AI model, such that it will allow impartial others to interrogate the AI model's accuracy. Which principle should win out in this situation?

The bioethicist Raanan Gillon, when commenting on Beauchamp and Childress in 1994, summarises this issue for us. Gillon's lesson is quite applicable to the extensive list of digital ethical principles we encountered earlier:

> [the principlist approach …] does not provide a method for choosing, which is a source of dissatisfaction to people who suppose that ethics merely comprises a set of ordered rules and that once the relevant information is fed into an algorithm or computer out will pop the answer ….[12]

However, Gillon argues that principlism can nonetheless provide a common set of moral commitments and moral language for us to think about ethical quandaries. In addition, we can seek to work out which principles are strongest or most compelling when they conflict in particular contexts.

Furthermore, in some cases we might be able to use critical thinking – or what Aristotle called practical wisdom – to work out ways of meeting the two or more principles that *seem* to clash but may in fact be reconcilable in some circumstances. In our simple analogical example, the New Year's resolution principles may be reconciled by, for instance, finding another way to exercise rather than going to the distant gym. Now recall the example of the company training an AI model on sensitive health data. Perhaps there will be a way of meeting, at least to some extent, both the principle of privacy and the principle of transparency in that sort of case. For example, perhaps the data can be kept relatively private by only sharing it with certain authorised experts tasked with checking the model, or by adequately de-identifying the data prior to sharing it more widely.

## Objection 3: Ethics Washing/Virtue Signalling

### Scenario 3: Looking good vs feeling good

Suppose you are diligently sticking to a gym routine, working out every week. To go along with it, you buy yourself some expensive sneakers and activewear which are very comfortable. To your surprise, people often compliment you over the choice of expensive sneakers you have. You obviously feel good!

You are now tempted to just go to the gym, do the bare minimum 5-minute jog on the treadmill, and just hang out at the gym's cafe instead. Sure enough, you get people talking to you quite often, even people whom you thought would never be interested in you.

This scenario very simply illustrates how you can just reap the 'reward' of visibility and social connections from the gym, without putting in the hard yards (which was your intention all along). Perhaps some of us are even guilty of wearing our running gear to work or to the shops to feel good about ourselves!

There is a parallel here with an objection to principlism vis-à-vis guidelines by organisations (especially tech companies): Accusations of such organisations merely performing *'virtue signalling'* or *'ethics washing'*.

Virtue signalling is an insincere espousal of values aimed at boosting the signaller's image and reputation. Preaching ethical principles can also serve the purpose of ethics washing. In ethics washing, an organisation pretends to be ethical but only to prevent other steps being taken to ensure its products and behaviour are indeed ethically justified. For example, voluntary guidelines setting out basic ethical principles might be produced to prevent the introduction of laws that could affect the organisation's interests in profit-making or in acquiring greater political power.

It is true that tech companies claiming to be responsible can behave hypocritically. It also seems fair to retain some scepticism about the sincerity of an organisation that is strongly driven by money and power. Indeed, we are now witnessing big tech companies amassing extreme wealth, dominating smaller competitors, and strongly influencing politics.

However, not all organisations that produce ethics guidelines are profit driven and not all are insincere. In addition, the ethical values that tech companies claim to practice may sometimes be used to hold them to account. For example, staff at Google – which had the memorable motto *'don't be evil'* – successfully protested its involvement in the Pentagon's *Project Maven*, which sought to use Google's expertise to improve weapons systems.[13] It is also the case that promoting ethical principles is consistent with changing the law. The widespread acceptance of appropriate ethical principles may even help in arguing in favour of certain legal restrictions on digital technology.

## Objection 4: Why Bother?

### Scenario 4: What's a New Year's Resolution worth anyway?

You catch up with some friends about your new resolution. To your surprise, two of them, Anna and Akshay, mention that they don't bother with making such resolutions. Puzzled, you ask them for more details. Anna explains that making resolutions every year is 'self-serving' and the hustle and bustle of daily life is actually doing us harm: Resolutions are only for us to comfort and distract ourselves. Akshay claims that resolutions to stay healthy and the like are 'luxuries' for us. However, Akshay's concern is towards others in circumstances who have more pressing matters to attend to, such as feeding their family, or carers' responsibilities, and how they won't appreciate the luxury of having time to entertain resolutions.

Analogously, this last objection to ethical principles and moral thinking is that they are 'beside the point' or a criticisable distraction from what really matters.

Firstly (as with the Anna analogy), the real problems with harmful digital technology, some argue, are its enabling social and power structures. For example, it may be suggested that the values and institutions of neoliberalism, unrestrained capitalism, and extreme individualism are responsible for tech's excesses. Indeed, the philosopher Ayn Rand, who argued for such values, is a role model for some Silicon Valley technologists.[14] In reply to this criticism, we need not deny that certain social and power structures enable unethical conduct from tech companies.

Secondly (as with the Akshay analogy), some object that application of principles, or even ethics, can interfere with other more pressing requirements. For example, ethics could plausibly be on the radar for big corporations or Western markets, since they have funding and staff to spare. Contrast that with, say, the small struggling technology vendor, or the solo app developer in the Global South or Majority World. Their priorities lie elsewhere, from acquiring funding to fixing technical issues, and it is not their fault that adhering to the checklist of principles is not high on their radar.

However, it does not follow from these objections that ethics or moral philosophy are irrelevant. For example, we may think that what is most wrong with, say, neoliberalist individualist thinking, is that it ignores justice and the promotion of the wellbeing and autonomy of other people, including the most vulnerable and least powerful.

Another way to think about it is that the other pressing concerns that prevent a struggling technology vendor from thinking about ethics first and foremost is due to a systemic imbalance of power and resources across the tech industry landscape. But this is also an ethical issue, one of, say, justice. Furthermore, there will likely be a range of ethical issues for those making technology to consider, beyond such hugely ambitious projects as changing the structure of society. Ethical thinking remains important, indeed vital, for addressing technologies and systems that impact heavily on others, even if we do need more than only thinking to make the world a better place.

* * *

In this chapter, we have looked at the potentials and pitfalls of principlism or of sets of principles in promoting ethical practice in the real world. A principlist-like approach in digital ethics (Chapter 5) looks, at face value, like an effective method in digital ethics – in the same way the four principles approach of Beauchamp and Childress was in biomedical ethics – augmented by supplementary principles specific to modern digital technologies.

However, as with our earlier explorations of ethical theories and frameworks, there are a few potential arguments against the practical efficacy of moral principles, including cherry-picking or vagueness of principles, actual and apparent conflict between principles, ethics-washing and virtue signalling, and the alleged irrelevance of principles in an unjust system.

Despite the four objections we posited, one could still be cautiously optimistic. Indeed, we still think that principlism is useful as a starting point in applying digital ethics to the use, appreciation, and design of technology. The relative simplicity and applicability of vital principles help technology stakeholders embrace ethical behaviour without the need to be philosophers.

There may be additional 'instrumental' reasons for adopting and promoting ethical principles. Organisations, for example, may endorse ethical principles partly to shore up their reputations and customer loyalty, at the same time as sincerely espousing those principles. Legal stakeholders might view adherence to digital ethics principles as part of meeting risk and compliance requirements of an organisation, seeing ethical principles as also being necessary to ensure the organisation does the right thing under the legal system.

In the following chapter, we round out our discussion of digital ethics by focusing on the unknown future, and entertaining ideas for continuing our digital ethics journey and future-proofing ourselves. In doing so, we offer practical advice and inspiration in helping you *transition to ethical technological practice*, in the spirit of this book and series.

## Notes

1. Floridi & Cowls (2019).
2. Jobin, Ienca & Vayena (2019).
3. IBM (n.d.) and Microsoft AI (n.d.).
4. DISR (2024, October 11).
5. ICRC (2024, November 18).
6. Jobin, Ienca, & Vayena (2019).
7. UNSCEB (n.d.).
8. OECD (n.d.).
9. IEEE (n.d.) and IEEE (2020)
10. Gotterbarn et al. (2018)
11. Beauchamp & Childress (2001).
12. Gillon (1994).
13. Shane & Wakabayashi (2018, April 4).
14. Bilton (2016, October 5).

## References

Beauchamp, T. L., & Childress, J. F. (2001). Principles of Biomedical Ethics. Oxford University Press.

Bilton, N. (2016, October 5). Silicon Valley's Most Disturbing Obsession. Vanity Fair. https://www.vanityfair.com/news/2016/10/silicon-valley-ayn-rand-obsession

DISR. (2024, October 11). Australia's AI Ethics Principles. Department of Industry, Science and Resources. https://www.industry.gov.au/publications/australias-artificial-intelligence-ethics-principles/australias-ai-ethics-principles

Floridi, L., & Cowls, J. (2019). A unified framework of five principles for AI in society. Harvard Data Science Review, 1(1). https://doi.org/10.1162/99608f92.8cd550d1

Gillon, R. (1994). Medical ethics: Four principles plus attention to scope. BMJ, 309(6948), 184. https://doi.org/10.1136/bmj.309.6948.184

Gotterbarn, D., Brinkman, B., Flick, C., Kirkpatrick, M. S., Miller, K., Varansky, K., & Wolf, M. J. (2018). ACM Code of Ethics and Professional Conduct. ACM. https://www.acm.org/code-of-ethics

IBM. (n.d.). AI Ethics. IBM. Retrieved May 18, 2025, from https://www.ibm.com/artificial-intelligence/ai-ethics

ICRC. (2024, November 18). BUILDING A RESPONSIBLE HUMANITARIAN APPROACH: THE ICRC'S POLICY ON ARTIFICIAL INTELLIGENCE. International Committee of the Red Cross. https://www.icrc.org/en/publication/building-responsible-humanitarian-approach-icrcs-policy-artificial-intelligence

IEEE. (n.d.). Ethics and Member Conduct Home. IEEE. Retrieved May 18, 2025, from https://www.ieee.org/about/ethics

IEEE. (2020). IEEE Code of Ethics. IEEE. https://www.ieee.org/about/corporate/governance/p7-8

Jobin, A., Ienca, M., & Vayena, E. (2019). The global landscape of AI ethics guidelines. Nature Machine Intelligence, 1(9), 389–399. https://doi.org/10.1038/s42256-019-0088-2

Microsoft AI. (n.d.). Responsible AI Principles and Approach. Retrieved May 18, 2025, from https://www.microsoft.com/en-us/ai/principles-and-approach

OECD. (n.d.). AI Principles. OECD. Retrieved May 18, 2025, from https://www.oecd.org/en/topics/ai-principles.html

Shane, S., & Wakabayashi, D. (2018, April 4). 'The Business of War': Google Employees Protest Work for the Pentagon. The New York Times. https://www.nytimes.com/2018/04/04/technology/google-letter-ceo-pentagon-project.html

UNSCEB. (n.d.). Principles for the Ethical Use of Artificial Intelligence in the United Nations System. UNSCEB. Retrieved May 18, 2025, from https://unsceb.org/principles-ethical-use-artificial-intelligence-united-nations-system

# 7

# *Moving Forward*

In this book, we started with the role of ethics and philosophy in technology and subsequently identified and categorised the issues surrounding contemporary tech (Chapters 1–3). We then looked at the ethical theories espoused by philosophers (Chapter 4), before taking a leaf from the domain of applied ethics with regards to digital ethical principles (Chapter 5).

The blossoming of digital technology in the 21st century has put us in a state of rapid and rather perplexing change. We even questioned the role of ethics given this turbulent situation (Chapter 6). It is, of course, difficult to know what social changes digital technologies will bring. Nonetheless, we can try to anticipate some possibilities and reflect on their ethics.[1]

## Future Possibilities

### From AI to AGI to Superintelligence?

Currently, the most dramatic form of new digital technology is Generative AI applications. These are already affecting many domains, including business, education, entertainment, social interaction, politics, and medicine. Image and audio generators and large language models (LLMs) are being used to execute tasks that very recently only humans could perform. This expansion of technology is likely to continue at pace.

One possible and perhaps likely development is in AI that increasingly undertakes more active than passive roles. AI systems could be increasingly used as 'autonomous agents'.[2] At present, much AI only responds to human prompting – think of asking an LLM a question and getting a reply. But an autonomous agent will be more capable, acting to varying degrees on their own. For example, they may carry out tasks such as analysing and replying to our emails without being asked to. These can be viewed as technical advances beyond simple computerised 'agents' or 'bots' we have nowadays.

These systems may also be used to automate cognitive activities in various workplaces, or at least to augment human-led activities.[3] In addition to performing intellectual work, AI might get better at being creative, such as generating interesting art, even if the results fall short of human creativity[4]

DOI: 10.1201/9781003393511-7

(Chapter 3). Robots may also become more autonomous and better able to intelligently act after sensing their environments. Such robots could potentially take on more and more roles that require both cognitive and physical skills.[5]

Increases in the abilities of digital technologies have created a vigorous debate about their potential and their limits. Some people think that we are on a pathway towards imminent **Artificial General Intelligence,** or **AGI.**[6] Roughly speaking, an AGI would have the broad cognitive capabilities of a human being. Such human-level AI would not just have abilities in one area, such as in chess-playing or language translation, but in many areas and for many problems that require human-style intelligence. Perhaps advances in quantum computing – based on properties of quantum physics – could enable extremely clever systems to be built.

A genuine AGI, being digital, would have certain advantages over humans and their biologically based intelligences. For instance, an AGI would have an enormous memory and an extraordinary reserve of knowledge. It would analyse, decide, and act at extreme speed. For such reasons, some people believe it is possible, or even likely, that an AGI would beget something far beyond human intelligence: an ultra or superintelligence.[7]

One idea runs like this. An AGI could, by means of its impressive skills, design an AI system that is a little more intelligent than itself. The resulting AI system could then design a more intelligent system – and so on. Proponents of this idea think that this process could result in a *'singularity'*, an intelligence explosion that quickly leads to a superintelligent AI.[8]

Such an AI would not only be incomprehensibly smart, but would also make changes in the world of its own accord. Even if the superintelligence initially lacks robotic features that permit direct physical activity, it could, some people think, trick human beings into performing various actions on its behalf, including facilitating its own spread. If the superintelligence is malign – or else benign but *not aligned* with human goals and needs – it might intentionally or inadvertently wreak havoc on our species, other species, and the planet itself.[9]

Clearly, the aforementioned sorts of possibilities raise big ethical issues. A key question here is whether we can address these issues with the toolkit of ethical ideas we have discussed in this book. After all, serious talk (as opposed to futuristic talk) of human-like robots, AGI, and superintelligence is very strange and new. These technologies sound like science fiction. Could it be that we need new ethical or philosophical thinking and approaches to deal with them?

## Revisiting Our Ethical Toolkit

There is reason to answer both *'yes and no'* to that question. Theories (Chapter 4), such as care ethics or utilitarianism, and principles (Chapter 5), such as beneficence, nonmaleficence, respect for autonomy, and justice, can still help us to morally steer through this unfamiliar terrain.[10] For example, human-level AI, if it eventuates, will still raise issues concerning benefit, harm, choice, and fairness. Issues such as changing the landscape of work,[11] and the democratic process, have been covered in Chapter 3. Other issues include the distribution of wealth, and even grander problems such as world hunger, world war, and climate change.

We can try to navigate such issues by carefully thinking about how to be beneficent, nonmaleficent, respectful of autonomy, and just. This could include deciding whether we should attempt to build or allow extremely powerful AI in the first place. These sorts of ethical principles will not vanish just because technology changes, even if the changes are considerable.

We could also look to insights from 2000-year-old virtue ethics[12] (Chapter 4). For example, virtues like humility about what we know and don't know could help us to avoid the vices of complacency and panic. As philosopher Shannon Vallor has argued, the virtue of practical wisdom could be particularly important. Being practically wise about emerging or possible digital technologies requires some practical knowledge and much thoughtful judgement. Take the example of trying to develop human-level AI. An initial question is whether AGI is even possible, and if so, when it might arise. Some people think human-level AI will probably arrive in the next few years or decades. This is suggested, they say, by recent spectacular advances in apparent comprehension and reasoning produced by dramatically scaled-up language models, which even (some contend) appear to show sparks of general intelligence.[13]

Others, however, are far more sceptical. Some feel that LLMs, despite their successes, are severely limited in their understanding and their reasoning ability by their mode of operation. LLMs, critics insist, merely find patterns in language that enable predictions of which words should be strung together in certain contexts.[14] Critics argue that the failure of LLMs to really comprehend or reason is demonstrated by their brittleness, including their tendency to *hallucinate* false information. Part of the problem may be that LLMs and other kinds of AI, lacking bodies, are cut off from the real world and unable to learn as humans and other animals do. All this, say critics, radically distinguishes human intelligence from current machine intelligence.

Critics also say that it is morally problematic to devote too much attention and resources to AGI and superintelligence – which are most likely problems for the far future or may never occur – especially if that comes at the expense of addressing existing ethical problems with more ordinary types of AI.[15] In contrast, those who think AGI is probable and perhaps imminent, and that superintelligent AI could follow in its wake, may divide into two camps.

One camp (the so-called AI '*boosters*') thinks or hopes that such AI will produce great benefits for humans that, with appropriate management, could outweigh the harms and injustices it might cause. For example, very smart AI might cure many diseases, extend lifespans, help address environmental destruction, and enhance education. This kind of AI is thus well worth pursuing. The other camp (the so-called AI '*doomers*') thinks or worries that such

AI could potentially cause mass unemployment, environmental devastation, and even human extinction, which of course outweighs any benefits it might deliver. We should therefore pause the development of AI when it becomes just a little more intelligent than it already is.[16] Alternatively, some people suggest we should contemplate the cognitive enhancement of humans – perhaps by integrating our brains with digital technology – so that we are not rendered utterly dim-witted alongside our ultra-clever creations.[17]

Knowing what to think about this dispute is not easy. A practically wise person seeking to understand it would certainly want to learn more about how AI and LLMs function to assess the prospects of AGI and superintelligence. They would also, in spite of the hype, carefully examine what various impartial experts – those who do not have financial or other conflicts of interest – have to say and whether there might be an emerging consensus one way or the other.

With that understanding, they might then do their best to work out whether we should worry about or strive to develop human-level AI, and, if so, how that might most ethically be done. For example, some people are currently thinking about how we can better match or align AI-based decisions and actions with human needs.[18] Others believe that specific AI regulation will be required to prevent major harms.[19]

Our familiar moral theories and principles should also help us out with such questions. They can, for example, remind us to carefully weigh possible human, animal, and environmental effects, both positive and negative, of pursuing or not pursuing more autonomous, smarter AI. If AI gets sufficiently powerful, we might need to work out whether AI systems *themselves* should be taught or programmed with certain moral theories and principles so that they are safer, beneficial, and respectful of people's choices and values.[20]

## Moral Agency of AI and Technological Systems?

On the other hand, such technology appears to raise some deep philosophical questions that may go beyond familiar ethical principles. Consider the question of whether some digital technologies themselves may come to have intrinsic moral value or status as humans and some or all animals do. This would mean that we don't just have ethical responsibilities *regarding* technologies, but ethical responsibilities *to* those technologies *for their own sake*.

Recognising that an AI system or a robot has ethical value or status, in its own right, would represent a radical change in our moral landscape. It would be reminiscent of science fiction tales, from *Star Wars* to *Blade Runner*, in which machines must be treated as part of the moral community and can be wronged or respected by us. This idea raises thorny philosophical issues.

One issue is what is required for qualifying as a member of the moral community. For example, is being conscious and sentient – including having the ability to feel pain and experience pleasure – sufficient for possessing a moral status that others must respect?[21] Does having moral status require other abilities such as self-consciousness and moral reasoning? Alternatively, could at least a minimal degree of moral status arise from a robot or AI having goals and autonomously and consistently pursuing them?[22] Or could moral status instead emerge from the ways we socially and subjectively interact with machines, not just from their 'ontological' nature or objective properties?[23]

If we do decide that future machines have (or at least *might* have) moral status, we would then need to further reflect on exactly how we should treat them. For example, may we ever *use* these entities for our purposes against their autonomous wills and preferences? When, if ever, might it be permissible to cause them 'distress'? Would they have a right to freely reproduce?

Interestingly, some philosophers are against the idea of creating machines that have moral status. They think robots and other machines should remain our tools or 'slaves'.[24] However, it is possible to argue that if machines can be sentient and experience happiness, we humans could even have an *obligation* to create them. A position of radical ethical '*long-termism*' even holds that we should create huge numbers of sentient machines, or perhaps human minds uploaded onto machines, to increase the amount of happiness in the universe over billions of years.[25] Some utilitarians might get on board with that project, though other thinkers might view it as foolish.

Another philosophical question is how we should relate to future machines even when they are not conscious or sentient. Some new technologies might increasingly resemble animals and humans and simulate friendly companions and confidants. The science fiction writer Philip K. Dick[26] imagined a world with synthetic humans and pets almost undisguisable from live ones. In his imaginary world, there was a definite preference for real humans and animals over artificial ones.

Yet, as we are already starting to see, some individuals like and even prefer the company of AI avatars which mimic friends or even romantic partners.[27] These personable simulacra are not accompanied by the uncertainties and risks present in human relationships. An important question is whether such interaction is beneficial or harmful, and to what degree we should facilitate, say, children's access to them? We could also ask whether being 'cruel' or 'sexist' to a machine, such as a robot or smart speaker with a female voice,[28] could make us more callous and prejudiced.[29]

Questions about character and risk are also relevant to possible advances in *virtual reality*. If technology advances and virtual reality becomes a more compelling experience than it is now, people may spend more and more time in it. This could include interacting with and 'befriending' virtual characters, as well as encountering other real people, or their preferred avatars, in the virtual space.

Philosophers ask tough questions about whether a virtual reality that gives the exact same experience as the ordinary world we live in – a scenario depicted in the popular film series *The Matrix* – is any the less 'real'.[30] After all, these philosophers argue, if our ordinary world turned out to be a digital simulation, we would surely not dismiss it as 'not real'.

However, we can also think about whether less futuristic but still vivid forms of virtual reality might create various ethical risks and concerns. For example, immersion in 'quasi-realistic' virtual spaces might still have the power to facilitate dangerous political radicalisation, social withdrawal, or addiction. Familiar moral theories and principles can surely come in handy here. Thus, understanding approaches such as deontology, virtue ethics, care ethics, utilitarianism, and principlism will continue to be important and relevant in our strange and changing world.

## Futureproofing Techniques

### Pre-mortems

One useful technique to aid ethical thinking about digital developments is known as the **pre-mortem**, which was developed by Gary Klein for evaluating projects.[31] Unlike post-mortems, which examine what went wrong after an event, pre-mortems are team exercises that anticipate what *could* go wrong with a project or plan. A digital ethics pre-mortem can be used to anticipate the negative outcomes of a digital technology before it is developed or implemented and to evaluate the ethics of doing so. The thought is that by imagining that the technology has already been deployed, and has resulted in severe ethical problems, we increase the chances of identifying its potential risks and thus of making better moral judgments about it. Think of it as a formal way of hypothesising potential effects of a piece of digital technology which could perhaps be a plotline for the TV series *Black Mirror*.[32]

A digital ethics pre-mortem exercise starts with individuals imagining that a given technology has been a complete failure due to ethical reasons. This first step involves private thinking about worst-case scenarios; it is done privately to ensure our brainstorming is not sidetracked by others' opinions. Individuals are encouraged to think widely and without inhibition about all the things that could go wrong. After this, individuals in the pre-mortem exercise come together as a team and share their thoughts, until as many sorts of ethical mishap as people can think of have been canvassed by the group.

For illustration, let's take the earlier example of AI companions.[33] Suppose a company decides to make an app containing custom-tailored AI romantic partners, modelled on celebrities that each user is personally attracted to. Suppose the AI – trained on likenesses of these celebrities – produce realistic personas that then 'fall in love' with users. Our digital ethics pre-mortem exercise requires us to freely imagine what has gone wrong sometime after the app or video game's release.

Possibilities of ethical disaster for this celebrity romance app are numerous. They could include users becoming heavily addicted to their virtual partners; people becoming increasingly socially isolated and relationships breaking down; increases in mental illness; celebrities feeling outraged by unauthorised or degrading use of their personas; legal cases against the company's use of deepfakes for living and deceased public figures; the stalking of real-life versions of virtual companions due to parasocial delusions; and AI avatars encouraging their human companions to do dangerous or anti-social acts (Chapter 2).

A possible *drawback* of pre-mortems is that we might exaggerate the negative implications of a technology, such as various possible harms, injustices, and infringements of autonomy that could hypothetically result, but that may or may not eventuate. At the same time, we might underestimate or ignore the positive ethical implications and benefits, both intended and unintended, of a new technology.

Nonetheless, the digital ethics pre-mortem exercise can be a helpful means of counterbalancing the hype and enthusiasm that sometimes greets new technologies. Pre-mortems can assist in ensuring that an intended technology is aligned with ethical principles like those we have considered. Those who design and deploy new digital technologies may therefore want to adopt this pre-emptive ethical technique.

## Diversifying Human Teams

Here's a quick thought experiment: Let's say you are taking part in a team-based general knowledge social trivia night. In the same vein as TV quiz shows, the questions are not easy and can come from a variety of subjects.

Given this, *what strategy would you employ in selecting your teammates*? Of course, it is in your best interest to maximise the skillset and specialties of your team. Stocking up the team with pop culture experts might not work if a question about Greek mythologies came up!

The same logic applies in the composition of team members working together in tech, be it in programming, designing a new robot, or introducing a new tech policy. Take, for instance, two examples of issues seen in Chapter 2: The unjustified confidence by technologists in the *Therac-25* machine (1980s) and AI recruitment tools susceptible to gender bias[34] (ca. 2010s). For the former, according to Sara Baase (2013), there were "…many indications that the makers and some users of the *Therac-25* were overconfident about the safety of the system".[35] Had a diversity of voices been included in the design and development of the system – such as a hospital physicist who painstakingly found the issues independent of the manufacturer (which actually happened in real life) – these problems might have been nipped in the bud.

As for the issues in gender-biased AI recruitment tools, teams should ideally not merely consist of technical staff (such as programmers and data scientists), but potentially also ethicists, sociologists or anthropologists, HR professionals, and legal experts. After all, the job market is not the exclusive domain of technologists; tools used in hiring will therefore need a plurality of stakeholders. Adding diversity to a team should not be an exercise in 'ethics washing' or simply to indicate 'virtue' and compliance (Chapter 6). In other words, the need for diversity stems from the following question: "How can... [technology companies and...] platforms design inclusive ethical and considerate guidelines when only certain people are brought to the table?".[36] Whether of a particular research/academic discipline, or of a certain demographic group, the inclusion of a limited set of voices will not enable a technological product to benefit all possible user groups (Chapter 3).

Research in academia and business, for example, have shown that diversity *does* improve outcomes. In academia, take the *Netflix Prize* – a competition in the 2000s to see if anyone could improve on the Netflix algorithm for predicting movie ratings – for the reward of $1 million. In developing this algorithm, computer science skills are not enough; competitors need to know how people from different walks of life would rank an assortment of movies. In other words, the more diverse a team is, the more perspectives, preferences, and cultural outlooks on movies they have, which will in turn make them better at making predictions:

> ...What had begun as a contest to determine the best data scientist became a demonstration of diversity bonuses... [requiring] knowledge about how a diverse consumer base rated [the] movies...[37]

From a statistical point of view, having people who are from different backgrounds helps drive innovation. Different people bring different viewpoints, and therefore different ideas. Combining lots of these ideas potentially makes serendipitous discovery of new ideas more likely.[38] For instance, the creation of the Velcro (hook-and-loop) fastener was inspired by engineer George de Mestral's strong curiosity about burdock plants in nature, which tends to stick to fabrics like hikers' socks. Here, biological observation influenced engineering. Different viewpoints also prevent overgeneralisation, groupthink, and confirmation bias. All these are important in real-world industry.[39]

We are increasingly aware[40] of the need for a plurality of viewpoints in our inquiry of digital ethics, beyond just the normative *Western, Educated, Industrialised, Rich and Democratic*[41] or WEIRD way of seeing things. We thus need to bring voices from the Global South or Majority World, LGBTIQA+ communities, ethnic minorities, people with disabilities, and other minoritised groups[42] to the table, before we can confidently claim that a piece of technology is truly inclusive.

## Beyond Regulations: Collective Action

As we have seen throughout this book, regulations and laws, such as the GDPR (in the EU), aim to protect people from some of the risks and harms of digital technology. However, there is no single set of laws that apply to everyone on the planet: Laws and regulations are jurisdiction dependent.

What is possible, irrespective of jurisdiction, is our ability to act collectively in the interest of the greater good. A good starting point is to be aware of the potential harms of technology (which you have already begun to do when you picked up this book) and to exercise your rights as an individual consumer. This can be as simple as adopting the considerations in Chapter 3, such as by being parsimonious with one's data when dealing with tech companies, or trimming down our social media footprints. It can also involve more technological measures, such as installing ad-blockers and privacy-preserving technologies.[43] If such small measures gain traction, it will signal to Big Tech that users really do care about their online wellbeing. One concrete example is in Apple's and Mozilla's push to create more privacy-respecting web browsers,[44] which would never have happened without consumer demand.

On a collective level, we can do many things to promote more ethical outcomes. As with the call to diversity discussed in the previous section, working together as a team can sometimes be preferable to working alone. This can take the form, for instance, of trade unions, which support members' interests when faced with issues arising from, say, technological disruption.[45] Here, we could enact the principle of solidarity we encountered in Chapter 5.

Communities of practice and supportive networks can also help the lone knowledge worker face ethical concerns and make it safer to speak up.[46] We also discussed in an earlier chapter how ethical puzzlement and confusion about the right thing to do, including how to resolve ethical dilemmas and conflicts of principles, can be aided by conversation with others, the formation of ethics committees or discussion groups, and the creation of suitable cultures in organisations for open moral dialogue.

With regard to data and code, new ways of collectivism have been inspired by traditional (offline) initiatives, for the public good. Take *data cooperatives*, or 'member-owned organisational structures that allow for their members to voluntarily pool data together and control it collectively for their mutual benefit, or the benefit of the wider community'.[47] Data cooperatives enable the users and adopters of tech such as ridesharing (Chapter 3) to have their say on their collective data.[48] Also, the open source movement (Chapter 2) and Creative Commons licensing scheme has led to communities benefiting from collaborative action, such as the development of the Linux operating system for the former (Chapter 2), and the curation and dissemination of knowledge via Wikipedia for the latter. These examples illustrate the power of collective action with regard to the asymmetric relationship between tech companies, their partners, and us, the data subjects.

Last but not least, we should not forget the role of whistleblowers who are to be commended for their role in bringing light to otherwise hidden but significant ethical issues in technology. Frances Haugen (Chapter 1),[49] Sarah Wynn-Williams,[50] Timnit Gebru,[51] and Peiter 'Mudge' Zatko[52] are examples of tech whistleblowers who have, at great personal sacrifice, exposed large ethical problems in tech companies, which are otherwise concealed from the outside world.[53] As a result of their actions, regulators, academics, and the public have gained a deeper understanding of the causes and effects of digital ethical issues that once were merely the subject of speculation.

## Accountability Methods for Technology

Another useful way of proactively anticipating future challenges is to build a practice of transparency and accountability about the technology products we build today. For the sake of argument, let's take the food we buy from the supermarket daily. If my doctor advised me to increase my fibre intake and decrease cholesterol, I would be on the lookout for the nutritional label on the food I purchase. Similarly, when I have a meal at the local restaurant, I will be attentive to allergen advice as well as the list of ingredients on the menu, if I have any dietary requirements or health concerns.

Given the aforementioned argument, it is perhaps surprising that such a uniform labelling mechanism does not exist for technological products. (Proposed initiatives and regulations such as the EU AI Act are steps in this direction, such as in labelling/assessing risk levels of specific AI applications.) However, technology deployers and creators can take a positive first step by way of ethical documentation practices, such as through *model cards* and *datasheets* for datasets.[54] Collectively, they "provide a standardised documentation for data-driven decision-making systems, and their corresponding datasets… [and] importantly, provide a starting point for a discussion on both existing (current, in-use) and future systems".[55]

One practical advantage that these practices have when compared to, say, having ethical guidelines (Chapter 5), is that they are designed to get stakeholders (creators, testers, maintainers, and decision-makers, amongst many) behind a piece of technology to reflect, document, and anticipate any ethical challenges in the process of documenting them. As with their food safety counterparts, these in turn enable end-users or adopters of technology to make an informed choice in their decision to use them. Again, the analogy of the nutritional/allergy labels for food comes to mind.

Also, when people are adversely impacted by decisions made by automated systems, concretely documenting the limitations of the systems has the effect of empowering these subjects to seek remedies and perhaps fair compensation if they have been wronged.[56]

Given that, there are some caveats[57] to these approaches that are worth outlining here. Firstly, as with sets of principles discussed in Chapter 5, these documentation practices are *voluntary and unenforceable* (as of time of writing). Secondly, there is nothing stopping from *'picking and choosing'* the things they choose to report on. If a particular aspect is unfavourable (such as an AI potentially causing harm to a certain group), this aspect could simply be left out from any documentation (Chapter 6). Thirdly, the *effort needed* in creating such documentation artefacts is not a trivial matter; this alone is a huge obstacle preventing uptake.

Hearteningly, these approaches are gaining traction in high-stakes areas such as in healthcare, where such transparency is crucial in maximising positive health outcomes for people.[58] We remain hopeful that these initiatives will gain traction the same way nutritional, allergy, and food content labels are the norm for food products.

## Digital Ethics Education

Some readers may have picked up this book because of an interest in gaining ethical knowledge and reasoning skills for their work with digital tech. This raises a question of whether ethics education might be beneficial or even essential for professionals who work with computers, AI, robotics, and so on. The need for an ethics understanding is reflected in ethical codes of professional practice for engineering and computing scientists, such as those from bodies such as the IEEE[59] and the ACM.[60] Accreditation bodies and professional organisations may also lay down requirements for ethics competency, education, and professionalism in engineering and IT degrees.

The increasing power and reach of digital technologies suggest that ethics education will need to become even more central to an education in these technologies.[61] This, we can now see, must go beyond professional ethics: Instead of seeing digital ethics as an 'add-on' for professionals, we should perhaps regard an understanding of digital ethics as fundamental or core knowledge. Digital ethics education could start during formative education when one first learns about computers and technology in school.[62] In tertiary education, this could take the form of standalone ethics subjects and/or as key learning objectives in standard curricula.[63] Continuing ethics education could also be made routine for working professionals, say, as an important step towards re-accreditation.

This emphasis would resemble the way that medical ethics is regarded as an indispensable component of education for health professionals.[64] Teaching health professionals about ethics is widely seen as essential. Part of the reason for this is that health professionals are in a position of power over their patients, who may suffer grave harm if not treated responsibly. There is thus good reason to ensure doctors and nurses have a keen awareness of ethical values and concepts. Another key reason is that since medical practice is complex, reasoning and behaving ethically is sometimes difficult.

These two features surely apply to contemporary digital technology. First, digital technologies, especially since the latest AI boom, have increasing power over human lives. As we have seen, humans, not least vulnerable and marginalised humans, can be very seriously impacted by digital technology in many ways. Digital technologies may also seriously affect nonhuman animals and the natural world (as seen in Chapters 2–3).

In addition, technology's growth in the last decade or two has made tech companies and their owners some of the richest and most powerful and influential people on earth. It seems important that those working for tech companies should have some understanding of digital ethics. In fact, creating ethically aware cultures in computing industries could be important for promoting more morally desirable technologies, especially when legislation and regulation lags behind tech development.

Secondly, the effects of digital technologies are complex, and morally reasoning about them can be difficult. As we have seen in this book, there are an almost unlimited number of possible digital devices and applications for all areas of human life.

Sceptics might object that digital ethics education is unnecessary, or even useless. Surely, they might say, people are either conscientious or not; learning ethical theory will not make them behave better. There is some truth to this. As Aristotle said, developing a good character that enables people to behave well requires practice and a genuine desire or commitment to do the right thing.

However, one also needs to know what good and bad behaviour is, and that means understanding ethical ideas and concepts. When the subject matter is complex, it also helps to see how ethical values and principles can clash, and to understand how we might navigate these moral conflicts. In that respect, learning about digital ethics seems to be useful, and increasingly vital, for 21st-century tech professionals.

* * *

Through this concluding chapter, we have presented a menu of future possibilities, based on the current discussions in the press and in academia. We have also presented a series of actionable strategies that digital practitioners can start adopting now to ensure digital ethics remains relevant across the tide of technological evolution.

As much as we'd like to continue exploring digital ethics with you, we must bring things to a close given the limited space we have in this book. We hope you have enjoyed the journey with us unravelling digital ethics, in line with the ethos of the *Transition to Tech* series.

However, we don't believe that this is the end. Instead, one could treat this introduction as the beginning of further exploration in digital ethics for years to come. This could be in your work or just for personal interest. We encourage you to explore the topics in this volume further, for example, by looking into the sources we have provided in our footnotes throughout this volume.

We conclude this book with these parting words, with the help of a panel of several large language models of today, reflecting on the lessons we've penned in these pages.

*It's essential to acknowledge the complexities and challenges of navigating the intersection of technology, business, and human values.*

*However, there is a beacon of hope in the form of individuals and organisations who choose to act with integrity, respect, and empathy.*

*The question is no longer whether we can build smarter technologies, but whether we can do so wisely and justly.*[65]

## Notes

1. Vallor (2018) and Vallor (2024).
2. McCorduck (2004).
3. Brynjolfsson, E. (2022).
4. Flick & Worrall (2022).
5. Coeckelbergh (2022) and Lin, Abney & Bekey (Eds.) (2014).
6. Goertzel (2007).
7. Good (1966).
8. Bostrom (2014).
9. Bostrom (n.d.).
10. To recap, ethical theories and representative references include: Care Ethics – Held (2006); Utilitarianism – de Lazari-Radek & Singer (2017); and Principlism – Beauchamp & DeGrazia (2004).
11. Suleyman & Bhaskar (2023).
12. Swanton (2003).
13. Bubeck et al. (2023).
14. Bender, Gebru, McMillan-Major & Shmitchell (2021).
15. Crawford (2021) and O'Neil (2016)
16. Pause Giant AI Experiments: An Open Letter (n.d.).
17. Bostrom (2014) and Kurzweil (2005).
18. Christian (2021) and Russell (2020).
19. Munn (2023).
20. Wallach & Asaro (Eds.) (2020).
21. A basic philosophical issue here, mentioned in Chapter 1, is how can we *know* whether a machine is, say, conscious or sentient.
22. Long et al. (2024).
23. Coeckelbergh (2010), Gordon & Gunkel (2022), and Gunkel (2012).
24. Bryson (2010).
25. See, e.g., discussions in Adams, Crary & Gruen (2023), Ord (2020), and Tegmark (2017).
26. Dick (1968).
27. Ciriello, Hannon, Chen & Vaast (2024).
28. Strengers & Kennedy (2020).
29. Coghlan, Vetere, Waycott & Barbosa Neves (2019) and Sparrow (2021).
30. Chalmers & Peacock (2022).
31. Klein (2007).
32. Many other exercises exist, which help developers anticipate challenges and prevent disasters related to tech. The examples here were inspired by Laura Summers's Ethics Litmus Tests, which we highly recommend. See Summers (n.d.).
33. Levy (2008).
34. Dastin (2018, October 11).
35. Baase (2013).
36. Bush et al. (2021, January 27).
37. Page (2017).
38. These ideas are thanks to David Green, a mentor, friend, and PhD supervisor of one of the authors. See Green (2004) and Green (2014).
39. Hewlett, Marshall & Sherbin (n.d.) and Hunt, Layton, & Prince (2015, January 1).

40. See Chapter 5 on diversifying the philosophical canon.
41. Henrich (2020).
42. Cheong et al. (2025), Kula et al. (2022), Liao & Huebner (2021), and Munn (2023).
43. Examples include uBlock Origin (https://github.com/gorhill/uBlock ) and EFF PrivacyBadger (https://privacybadger.org/).
44. See, e.g., https://www.apple.com/safari/privacy/ and https://support.mozilla.org/en-US/kb/recommended-firefox-privacy-and-security-settings-families.
45. Aussie big tech unionisation a sign of AI times (2025, May 26).
46. Addressing Ethical Issues as a Tech Worker (n.d.).
47. Hardjono & Pentland (2019) and Petreski & Cheong (2024).
48. Petreski & Cheong (2024).
49. Haugen (2023).
50. Wynn-Williams (2025).
51. Hao (2020, December 4).
52. Kleinman & Clayton (2022, September 13).
53. Bhuiyan (2021, October 9) and The Tech Worker Handbook (n.d.).
54. Chan (2023), Gebru et al. (2021), and Mitchell et al. (2019).
55. Cheong (2024).
56. Venkatasubramanian & Alfano (2020).
57. Cheong (2024) and Rosenstock (2023, May 19)
58. Rostamzadeh et al. (2022)
59. IEEE Code of Ethics (n.d.)
60. ACM Code 2018 Task Force (2018).
61. Paltiel, Cheong, Coghlan & Lederman (2023).
62. Such initiatives include, e.g., the Girl Geek Academy initiative (https://girlgeekacademy.com/)
63. Oliveira (n.d.).
64. Andersson et al. (2022).
65. Adapted from the replies of llama3.2:3b, mistral-0.3-7b, and ChatGPT (GPT-4o), respectively (knowledge-bases as of 11 June 2025), to the prompt, "How to conclude a book on digital ethics with an example".

## References

ACM Code 2018 Task Force. (2018). ACM Code of Ethics and Professional Conduct. ACM. https://www.acm.org/code-of-ethics

Adams, C. J., Crary, A., & Gruen, L. (2023). The Good It Promises, the Harm It Does: Critical Essays on Effective Altruism. Oxford University Press.

Addressing Ethical Issues as a Tech Worker. (n.d.). Mozilla Foundation. Retrieved June 17, 2025, from https://www.mozillafoundation.org/en/initiatives/great-tech-great-responsibility/addressing-ethical-issues-tech-worker/

Andersson, H., Svensson, A., Frank, C., Rantala, A., Holmberg, M., & Bremer, A. (2022). Ethics education to support ethical competence learning in healthcare: An integrative systematic review. BMC Medical Ethics, 23(1), 29. https://doi.org/10.1186/s12910-022-00766-z

Aussie Big Tech Unionisation a Sign of AI Times. (2025, May 26). Australian Financial Review. https://www.afr.com/technology/aussie-big-tech-unionisation-a-sign-of-ai-times-20250526-p5m28o

Baase, S. (2013). 8.2 Case Study: The Therac-25. In A Gift of Fire: Social, Legal, and Ethical Issues for Computing Technology (4th ed.). Pearson.

Beauchamp, T. L., & DeGrazia, D. (2004). Principles and principlism. In Handbook of Bioethics: Taking Stock of the Field from a Philosophical Perspective (1st ed.). Imprint: Springer. https://doi.org/10.1007/1-4020-2127-5

Bender, E. M., Gebru, T., McMillan-Major, A., & Shmitchell, S. (2021). On the dangers of stochastic parrots: Can language models be too big? 🦜. Proceedings of the 2021 ACM Conference on Fairness, Accountability, and Transparency, 610–623. https://doi.org/10.1145/3442188.3445922

Bhuiyan, J. (2021, October 9). 'Welcome to the Party': Five Past Tech Whistleblowers on the Pitfalls of Speaking Out. The Guardian. https://www.theguardian.com/technology/2021/oct/08/tech-whistleblowers-facebook-frances-haugen-amazon-google-pinterest

Bostrom, N. (n.d.). Ethical Issues in Advanced Artificial Intelligence. Retrieved June 17, 2025, from https://philarchive.org/rec/BOSEII

Bostrom, N. (2014). Superintelligence: Paths, Dangers, Strategies. Oxford University Press.

Brynjolfsson, E. (2022). The turing trap: The promise & peril of human-like artificial intelligence. Daedalus, 151(2), 272–287. https://doi.org/10.1162/daed_a_01915

Bryson, J. J. (2010). Robots should be slaves. In Y. Wilks (Ed.), Close Engagements With Artificial Companions: Key Social, Psychological, Ethical and Design Issues (pp. 63–74). John Benjamins Publishing Company. https://www.degruyterbrill.com/document/doi/10.1075/nlp.8.11bry/html

Bubeck, S., Chandrasekaran, V., Eldan, R., Gehrke, J., Horvitz, E., Kamar, E., Lee, P., Lee, Y. T., Li, Y., Lundberg, S., Nori, H., Palangi, H., Ribeiro, M. T., & Zhang, Y. (2023). Sparks of Artificial General Intelligence: Early Experiments with GPT-4 (No. arXiv:2303.12712). arXiv. https://doi.org/10.48550/arXiv.2303.12712

Bush, G., Nadeem, M., Cheong, M., Leins, K., & Coghlan, S. (2021, January 27). Trump, Nipples and the Hypocrisy of the Social Media Giants. Pursuit. https://pursuit.unimelb.edu.au/articles/trump-nipples-and-the-hypocrisy-of-the-social-media-giants

Chalmers, D. J., & Peacock, T. (2022). Reality+: Virtual Worlds and the Problem of Philosophy. Allen Lane, an imprint of Penguin Books.

Chan, A. (2023). GPT-3 and InstructGPT: Technological Dystopianism, Utopianism, and "Contextual" Perspectives in AI Ethics and Industry. AI and Ethics, 3(1), 53–64. https://doi.org/10.1007/s43681-022-00148-6

Cheong, M. (2024). Data Equity = Health Equity: Desiderata for Auditing of Biases in Current Data-driven Health Systems [Unpublished manuscript].

Cheong, M., Abuzo, S., Hata, H., Kevin, P., Kula, W., Mirou, B., Treude, C., Wang, D., & Kula, R. G. (2025). Building Bridges Across Papua New Guinea's Digital Divide in Growing the ICT Industry (No. arXiv:2501.09482). arXiv. https://doi.org/10.48550/arXiv.2501.09482

Christian, B. (2021). The Alignment Problem: Machine Learning and Human Values (First published as a Norton paperback). W. W. Norton & Company.

Ciriello, R., Hannon, O., Chen, A. Y., & Vaast, E. (2024). Ethical tensions in human-AI companionship: A dialectical inquiry into replika. Hawaii International Conference on System Sciences. https://doi.org/10.24251/HICSS.2024.058

Coeckelbergh, M. (2010). Robot rights? Towards a social-relational justification of moral consideration. Ethics and Information Technology, 12(3), 209–221. https://doi.org/10.1007/s10676-010-9235-5

Coeckelbergh, M. (2022). Robot Ethics. The MIT Press.

Coghlan, S., Vetere, F., Waycott, J., & Barbosa Neves, B. (2019). Could social robots make us kinder or crueller to humans and animals? International Journal of Social Robotics, 11(5), 741–751. https://doi.org/10.1007/s12369-019-00583-2

Crawford, K. (2021). Atlas of AI: Power, Politics, and the Planetary Costs of Artificial Intelligence. Yale University Press.

Dastin, J. (2018, October 11). Insight – Amazon Scraps Secret AI Recruiting Tool that Showed Bias Against Women | Reuters. Reuters. https://www.reuters.com/article/world/insight-amazon-scraps-secret-ai-recruiting-tool-that-showed-bias-against-women-idUSKCN1MK0AG/

de Lazari-Radek, K., & Singer, P. (2017). Utilitarianism: A Very Short Introduction. Oxford University Press.

Dick, P. K. (1968). Do Androids Dream of Electric Sheep? Doubleday.

Flick, C., & Worrall, K. (2022). The ethics of creative AI. In C. Vear, & F. Poltronieri (Eds.), The Language of Creative AI: Practices, Aesthetics and Structures (pp. 73–91). Springer International Publishing. https://doi.org/10.1007/978-3-031-10960-7_5

Gebru, T., Morgenstern, J., Vecchione, B., Vaughan, J. W., Wallach, H., Daumé, H., III, & Crawford, K. (2021). Datasheets for datasets. Communications of the ACM, 64(12), 86–92. https://doi.org/10.1145/3458723

Goertzel, B. (2007). Artificial General Intelligence (C. Pennachin, Ed.). Springer.

Good, I. J. (1966). Speculations concerning the first ultraintelligent Machine*. In F. L. Alt & M. Rubinoff (Eds.), Advances in Computers (Vol. 6, pp. 31–88). Elsevier. https://doi.org/10.1016/S0065-2458(08)60418-0

Gordon, J.-S., & Gunkel, D. J. (2022). Moral status and intelligent robots. The Southern Journal of Philosophy, 60(1), 88–117. https://doi.org/10.1111/sjp.12450

Green, D. G. (2004). Serendipity Machine: A Voyage of Discovery Through the Unexpected World of Computers. Allen & Unwin.

Green, D. G. (2014). Of Ants and Men: The Unexpected Side Effects of Complexity in Society (1st ed.). Springer Berlin Heidelberg : Imprint: Copernicus. https://doi.org/10.1007/978-3-642-55230-4

Gunkel, D. J. (2012). The Machine Question: Critical Perspectives on AI, Robots, and Ethics. MIT Press.

Hao, K. (2020, December 4). We Read the Paper that Forced Timnit Gebru Out of Google. Here's What It Says. MIT Technology Review. https://www.technologyreview.com/2020/12/04/1013294/google-ai-ethics-research-paper-forced-out-timnit-gebru/

Hardjono, T., & Pentland, A. (2019). Data Cooperatives: Towards a Foundation for Decentralized Personal Data Management (No. arXiv:1905.08819). arXiv. https://doi.org/10.48550/arXiv.1905.08819

Haugen, F. (2023). The Power of One: How I Found the Strength to Tell the Truth and Why I Blew the Whistle on Facebook (1st ed.). Little, Brown and Company.

Held (2006). The Ethics of Care: Personal, Political, and Global. Oxford University Press.

Henrich, J. (2020). The WEIRDest People in the World: How the West Became Psychologically Peculiar and Particularly Prosperous (1st ed.). Farrar, Straus and Giroux.

Hewlett, S. A., Marshall, M., & Sherbin, L. (n.d.). How Diversity Can Drive Innovation. Harvard Business Review. Retrieved June 17, 2025, from https://hbr.org/2013/12/how-diversity-can-drive-innovation

Hunt, V., Layton, D., & Prince, S. (2015, January 1). Why Diversity Matters [McKinsey & co.]. https://www.mckinsey.com/capabilities/people-and-organizational-performance/our-insights/why-diversity-matters

IEEE Code of Ethics. (n.d.). IEEE. Retrieved June 17, 2025, from https://www.ieee.org/about/corporate/governance/p7-8

Klein, G. (2007). Performing a Project Premortem. Harvard Business Review, September. https://hbr.org/2007/09/performing-a-project-premortem

Kleinman, Z., & Clayton, J. (2022, September 13). Twitter Misleading the Public, Whistleblower Says. BBC. https://www.bbc.com/news/technology-62889754

Kula, R. G., Treude, C., Hata, H., Baltes, S., Steinmacher, I., Gerosa, M. A., & Amini, W. K. (2022). Challenges for inclusion in software engineering: The case of the emerging Papua New Guinean Society. IEEE Software, 39(3), 67–76. https://doi.org/10.1109/MS.2021.3098116

Kurzweil, R. (2005). The Singularity Is Near: When Humans Transcend Biology. Penguin Publishing Group.

Levy, D. N. L. (2008). Love and Sex With Robots: The Evolution of Human-Robot Relationships (1st ed). HarperCollins Publishers.

Liao, S., & Huebner, B. (2021). Oppressive things. Philosophy and Phenomenological Research, 103(1), 92–113. https://doi.org/10.1111/phpr.12701

Lin, P., Abney, K., & Bekey, G. A. (Eds.). (2014). Robot Ethics: The Ethical and Social Implications of Robotics (First MIT Press paperback edition). The MIT Press.

Long, R., Sebo, J., Butlin, P., Finlinson, K., Fish, K., Harding, J., Pfau, J., Sims, T., Birch, J., & Chalmers, D. (2024). Taking AI Welfare Seriously (No. arXiv:2411.00986). arXiv. https://doi.org/10.48550/arXiv.2411.00986

McCorduck, P. (2004). Machines Who Think: A Personal Inquiry into the History and Prospects of Artificial Intelligence (2nd ed.). A K Peters.

Mitchell, M., Wu, S., Zaldivar, A., Barnes, P., Vasserman, L., Hutchinson, B., Spitzer, E., Raji, I. D., & Gebru, T. (2019). Model cards for model reporting. Proceedings of the Conference on Fairness, Accountability, and Transparency, 220–229. https://doi.org/10.1145/3287560.3287596

Munn, L. (2023). The uselessness of AI ethics. AI and Ethics, 3(3), 869–877. https://doi.org/10.1007/s43681-022-00209-w

O'Neil, C. (2016). Weapons of Math Destruction: How Big Data Increases Inequality and Threatens Democracy. Penguin UK.

Oliveira, E. (n.d.). Generative AI – Project-Based Course Notes. Retrieved June 17, 2025, from https://cis-projects.github.io/project_based_course_notes/extras/genai.html

Ord, T. (2020). The Precipice: Existential Risk and the Future of Humanity. Hachette Books.

Page, S. E. (2017). The Diversity Bonus: How Great Teams Pay Off in the Knowledge Economy. Princeton University Press. https://doi.org/10.2307/j.ctvc77c0h

Paltiel, M., Cheong, M., Coghlan, S., & Lederman, R. (2023). Approaches and models for teaching digital ethics in information systems courses – A review of the literature. Australasian Journal of Information Systems, 27. https://doi.org/10.3127/ajis.v27i0.4517

Pause Giant AI Experiments: An Open Letter. (n.d.). Future of Life Institute. Retrieved June 14, 2025, from https://futureoflife.org/open-letter/pause-giant-ai-experiments/

Petreski, D., & Cheong, M. (2024). Data Cooperatives: A Conceptual Review. ICIS 2024 Proceedings. https://aisel.aisnet.org/icis2024/lit_review/lit_review/15

Rosenstock, S. (2023, May 19). A House of Cards for LLM Transparency. ChatLLM 23, University of Sydney. https://www.youtube.com/watch?v=tKiQgAR0I_Y

Rostamzadeh, N., Mincu, D., Roy, S., Smart, A., Wilcox, L., Pushkarna, M., Schrouff, J., Amironesei, R., Moorosi, N., & Heller, K. (2022). Healthsheet: Development of a transparency artifact for health datasets. Proceedings of the 2022 ACM Conference on Fairness, Accountability, and Transparency, 1943–1961. https://doi.org/10.1145/3531146.3533239

Russell, S. J. (2020) Human Compatible: Artificial Intelligence and the Problem of Control. Penguin Books.

Sparrow, R. (2021). Virtue and vice in our relationships with robots: Is there an asymmetry and how might it be explained? International Journal of Social Robotics, 13(1), 23–29. https://doi.org/10.1007/s12369-020-00631-2

Strengers, Y., & Kennedy, J. (2020). The Smart Wife: Why Siri, Alexa, and Other Smart Home Devices Need a Feminist Reboot. The MIT Press.

Suleyman, M., & Bhaskar, M. (2023). The Coming Wave: Technology, Power, and the Twenty-First Century's Greatest Dilemma. Crown.

Summers, L. (n.d.). Ethics Litmus Tests. Ethics Litmus Tests. Retrieved June 14, 2025, from https://www.ethical-litmus.site

Swanton, C. (2003). Virtue Ethics: A Pluralistic View. Oxford University Press.

Tegmark, M. (2017). Life 3.0: Being Human in the Age of Artificial Intelligence. Alfred A. Knopf.

The Tech Worker Handbook. (n.d.). Retrieved June 17, 2025, from https://techworkerhandbook.org/

Vallor, S. (2018). Technology and the Virtues: A Philosophical Guide to a Future Worth Wanting (First Issued as an Oxford University Press Paperback). Oxford University Press.

Vallor, S. (2024). The AI Mirror: How to Reclaim Our Humanity in the Age of Machine Thinking. Oxford University Press.

Venkatasubramanian, S., & Alfano, M. (2020). The philosophical basis of algorithmic recourse. Proceedings of the 2020 Conference on Fairness, Accountability, and Transparency, 284–293. https://doi.org/10.1145/3351095.3372876

Wallach, W., & Asaro, P. (Eds.). (2020). Machine Ethics and Robot Ethics. Routledge. https://doi.org/10.4324/9781003074991

Wynn-Williams, S. (2025). Careless People: A Story of Where I Used to Work. Power. Greed. Madness. Macmillan.

# *Index*

Made in the USA
Monee, IL
12 November 2026

e8e5b7b7-04ae-4bce-871e-3b1240ea678bR01